I USED TO KNOW THAT MATHS

π

I USED TO KNOW THAT
MATHS

STUFF YOU FORGOT FROM SCHOOL

CHRIS WARING

FOREWORD BY CAROLINE TAGGART

This paperback edition first published in 2014

First published in Great Britain in 2010 by
Michael O'Mara Books Limited
9 Lion Yard
Tremadoc Road
London SW4 7NQ

A CIP catalogue record for this book is available from the British Library.

Papers used by Michael O'Mara Books Limited are natural, recyclable products made from wood grown in sustainable forests. The manufacturing processes conform to the environmental regulations of the country of origin.

ISBN: 978-1-84317-474-5 in hardback print format
ISBN: 978-1-78243-255-5 in paperback print format
ISBN: 978-1-84317-933-7 in ePub format
ISBN: 978-1-84317-934-4 in Mobipocket format

3 4 5 6 7 8 9 10

www.mombooks.com

Typeset by www.glensaville.com

Printed and bound by CPI Group (UK) Ltd, Croydon, CR0 4YY

CONTENTS

FOREWORD

When the original version of *I Used to Know That* was published, I spent a very jolly couple of days in a small BBC studio in central London. With headphones over my ears and a microphone in front of me, I talked to people on radio stations all over the country about the book: why I had written it, what they liked about it and what brought back hideous memories.

To my surprise, the hideous memories were what excited people most. Top of the list – and this bit wasn't a surprise – was maths. One listener said that just looking at the letters a+b=c on the page had brought him out in a cold sweat, even though he no longer had any idea why. Another radio station carried out a series of interviews in the street asking people, among other things, if they knew who Pythagoras was. 'Oh yes,' said one man, 'he's to do with triangles and angles and all that malarkey.'

I thought that was wonderful: 'all that malarkey' summed up perfectly the way many of my generation were taught. We had to learn it (whatever 'it' was); we were never really told why; and, once exams were over, unless we went on to be engineers or historians, we never thought about it again. But it lingered somewhere at the back of our minds, which may be why *I Used to Know That* touched a chord.

However, covering five major subjects and including a catch-all chapter called 'General Studies' meant that a single small volume couldn't hope to deal with anything in much depth. This is where the individual titles in this series come in; if *I Used to Know That* reminded us of things that we learnt once, these books will expand on them and explain why they were important. If you enjoy this one, look out for *I Used to Know That: English, Geography, General Science* and *History* as well.

The teaching of maths in my youth was perhaps the prime example (worst culprit?) in the 'never really told why' category: OK, a+b=c (or, on a bad day, $a^2+b^2=c^2$) and, yes indeed, the square on the hypotenuse equals the sum of the squares on the other two sides, but who cares? Why does it matter? These are questions that Chris Waring sets out to answer, with examples ranging from a ladder propped against a wall to the probability of being dealt a Royal Straight Flush in poker. If only they had taught us *that* when we were fourteen, life might have turned out rather differently.

On a less controversial level, Chris reminds us of the basics of multiplying and dividing, fractions and decimals, means and medians, and how to calculate the area of a circle or construct a tetrahedron. (And if you have forgotten what a tetrahedron is, he reminds you of that, too.) Every page of this reader-friendly book will bring back fond or not-so-fond memories of equations and percentages, cosines and pie charts – or of being kept in after school and made to write out 100 times, 'I must not stare out of the window during maths lessons.'

I imagine that the gentleman who came out in a cold sweat at the thought of algebra will find this less than soothing; for the rest of us, *I Used to Know That: Maths* explains the unexplained or the long forgotten – and reminds us of the mathematical reason why we have never (yet) won the lottery.

Caroline Taggart

INTRODUCTION

Maths is one of the only school subjects that almost everybody looks back on with a mixture of dread and frustration. Worst of all, there was invariably some unassuming kid at the front of the class who could do it effortlessly, which just added to the annoyance. With the compulsory O level or GCSE out of the way, most people relegate maths to the backs of their minds, hoping never to hear the phrases 'long division', 'negative correlation' or – horror of horrors – 'quadratic equations' ever again. They certainly don't want to be prompted to find the value of x.

When maths is presented to you in terms of hypothetical triangles and pointless probability (as if you had nothing better to do than roll a dice 100 times), it's easy to think of it as being entirely useless for life in the real world. But as a maths teacher, I make it my business to try to communicate the *purpose* of maths as well as all the rules. From keeping score in darts to building a skyscraper, maths does indeed have a purpose.

Maths is ordered, logical and, at times, philosophical. Some might even say it's beautiful. Maths is the language that is used by scientists to describe and explore our physical universe. Maths is problem-solving using logic, which is why employers in all fields are keen to recruit people with a good maths qualification. Maths is useful – almost every modern item, from a train station to a mobile phone, operates on principles that are fundamentally governed by mathematics.

My aim in writing this book is to start with the basics and gradually work through each topic so that it actually makes sense. For the most part, I'll be jogging your memory rather than embarking upon something you've never heard of. The further you read, the more I hope you'll start to feel like those maths geeks at the front of the class: instinctively able to apply maths to practical situations.

So whether you are reading this so you can help your son or daughter with their maths homework or because you've simply decided that it can't really have been that impossible after all, it is my fond hope that you will start to enjoy maths and share the satisfaction that comes from not only getting the answer right – but also being able to prove it.

ARITHMETIC

Arithmetic is the basic, everyday manipulation of numbers. Every time you work out how long it is until you finish work, check your change or guiltily calculate how many units of alcohol you have consumed this week, you are doing arithmetic.

To a mathematician, arithmetic is where you perform an operation on two or more numbers. These operations are commonly called addition, subtraction, multiplication and division, and arithmetic as a whole falls into two categories: the stuff you do in your head or on your fingers – mental arithmetic – and the stuff you did on paper at school but now do on a calculator, which we'll call paper arithmetic.

TEN DIGITS

Count yourself very fortunate that, way back when, someone came up with the ten numerals – 0, 1, 2, 3, 4, 5, 6, 7, 8, 9 – and what we call the decimal system. Why ten? How many fingers am I holding up?

Having ten fingers, or digits (aha!), means we humans, who learn to count and do basic arithmetic on our fingers, are happiest thinking about numbers in lots of ten. Roman numerals, which we had previously, are completely useless for doing paper arithmetic. Working out in which year your favourite costume drama was produced is hard enough – can you imagine trying to do long division with those numbers?

MENTAL ARITHMETIC

Good mental arithmetic takes practice. Not the sort of practice where you sit down and just do some mental arithmetic for a bit. We've all got much better things to do. Mental arithmetic is the sort of practical maths you do every day without realizing it: how much is your share of the bill? How long until the next train? How late for work will you be if you miss it?

We all have reasonable mental arithmetic when it comes to straightforward adding and subtracting. What we tend to be *less* reasonable at, however, is multiplication and division, which are flip sides of the same coin.

Times Tables Tricks and Cheats

If the phrase 'times tables', especially when used in conjunction with the word 'test', gives you a nightmarish flashback to ritual humiliation at school, you are not alone. Unfortunately, though, our paper methods for multiplication and division (pp.20–32) are no use for numbers lower than 10, so the only thing for it is to learn the damned times table.

The basic table has 144 things to learn on it, which seems rather excessive. But if I delete all the repetition (5×3 is the same as 3×5) and assume you are pretty handy with your 1 times table (i.e. you can count), you're left with this:

_	1	2	3	4	5	6	7	8	9	10	11	12
1												
2		4	6	8	10	12	14	16	18	20	22	24
3			9	12	15	18	21	24	27	30	33	36
4				16	20	24	28	32	36	40	44	48
5					25	30	35	40	45	50	55	60
6						36	42	48	54	60	66	72
7							49	56	63	70	77	84
8								64	72	80	88	96
9									81	90	99	108
10										100	110	120
11											121	132
12												144

This leaves you with a mere 66 things to learn, less than half of the whole table. And let's face it: if darts players can have good mental arithmetic, so can you.
Here are some tricks:

- For the **2 times table**, just double the number. So 2×8 is double 8, which is 16.
- For the **4 times table**, double the number twice. So 4×8 is double double 8, which is double 16, which is 32.
- For the **8 times table**, double the number three times. So 8×8 is double double double 8, which is double double 16, which is double 32, which is 64.
- For the **10 times table**, add a 0. So 2×10 is 2 with a 0 on the end: 20.
- For the **5 times table**, multiply by 10, and halve the result. So 6×5 is half of 6×10 (60), which is 30.
- For the **11 times table**, just repeat the number you're multiplying by. For instance, 8×11 is 88 and 4×11 is 44. This only works up to 9×11 (99), but you already know that 10×11 is 11 with a 0 on the end – 110 – so you only have to learn 11×11=121 and 12×11=132, and you'll have the whole thing memorized.
- For the **9 times table**, hold your hands out in front

of you with your fingers out. Counting from the left, fold down the finger that corresponds to the number of nines you want. So for 4×9, you fold down the fourth finger from the left, which is the index finger of your left hand:

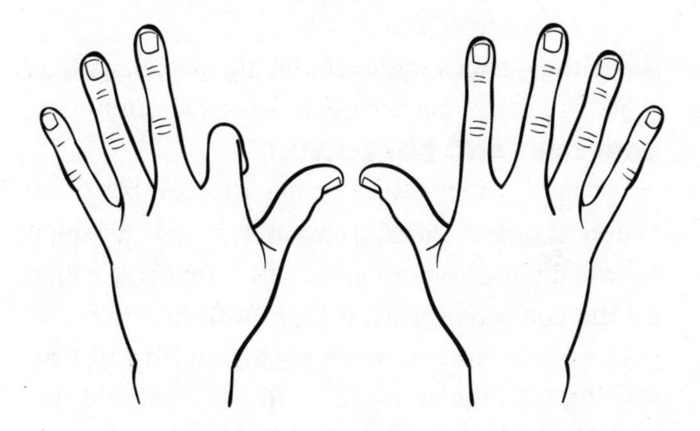

This leaves you with three fingers on the left and six on the right, which tells you that 4×9=36. Likewise, for 8×9, you'd put down the middle finger of your right hand to work out that your answer is 72. This only works up to 10×9, but there's a trick above for 11×9 (99), so you only need to memorize 12×9=108.

PAPER ARITHMETIC

Adding and subtracting are pretty straightforward and I won't go into them too much here. What tends to throw people of all ages is multiplication and division, particularly of the 'long' variety...

CARRYING AND BORROWING

When it comes to adding and subtracting on paper, we usually use column arithmetic. That is, we line up the columns (units, tens, hundreds, etc.) and perform the operation on each column in turn, starting with the units, filling in our answer at the bottom and then working our way left.

But what happens if the total of one of our column calculations is 10 or more? Where does the extra digit go? This is where carrying comes in. The joy – that's right, joy – of the decimal system is that if you end up with 10 of one column, you can simply add 1 to the column on the left, since 10 units amount to 1 ten, and 10 tens amount to 1 hundred, and so on. If our units column ends up with a ten in it, we simply move – or carry – it into the tens column.

When I was at school, carries were often put in as tiny little digits, quite often beneath where the answer was being written. This can end up looking a right old mess, so I'd recommend using whopping great carries at the top of the sum. Take 189+93, in which you have 9+3=12 and 1+8+9=18, and consequently have to carry two 1s:

$$\begin{array}{cccc} & 1 & 1 & \\ & 1 & 8 & 9 \\ + & & 9 & 3 \\ \hline & 2 & 8 & 2 \end{array}$$

Borrowing is basically carrying backwards, and you do it in subtraction when the number you start with is smaller than the number you need to subtract. Let's take 485–327. The first calculation gives you 5–7=-2, and things will rapidly go pear-shaped if I start inserting negatives into the middle of a number. What I *can* do to make sense of the sum is borrow 10 from the tens column, to turn 5 into 15. Taking 10 away from the 5's neighbour, 8, is easier than it looks because the 8 is in the tens column and is actually worth 80. The 8 loses 10 and becomes a 7, and the 5 gains 10 to become 15, and the sum is suddenly much more manageable:

$$\begin{array}{cccc} & 4 & 7\not{8} & {}^{1}5 \\ - & 3 & 2 & 7 \\ \hline & 1 & 5 & 8 \end{array}$$

If you ever need to borrow from a 0, you can simply turn the 0 into a 9 – thank heavens once again for the decimal system – but then you'd also need to borrow from the 'proper' number next to 0 so as not to alter your original number. (Take 200–73, for instance: if you turned the second 0 into a 10 and the first 0 into a 9, the 200 would be transformed into 290+10. By borrowing from the 2 as well, the 200 becomes 190+10, i.e. 200.)

If the number next to the 0 is another 0, you just have to keep on going until you hit a 'proper' number:

```
   4   9   9   9
   5  10  10  10  14
 -     9   8   7   6
 -------------------
   4   0   1   2   8
```

Multiplication

Multiplication is a way of doing lots of adding quickly. If I wanted to know what 8×17 is, I could add together eight 17s or seventeen 8s. This is quite a tedious process, however, so multiplying uses some basic number facts to take a shortcut.

To work out 8×17, I can break the 17 down into more manageable chunks. Armed with my times tables and the knowledge that 17=10+7, I can now simply work out 8×10 (80) and 8×7 (56) and add the answers together (136).

This, it turns out, is exactly what multiplication sums on paper let us do: break down any multiplication sum, no matter how hard, into a series of easier multiplication sums that we can add together to find the total.

In my years of teaching, I've come across three really handy methods for doing multiplication. I'll go through all three and you can choose whichever one works best for you. For each method, we'll work out the same sum – 143×25 – to make it easier to compare.

Splitting up the numbers, we can see that 143=100+40+3 and that 25=20+5, so at some point in each method I'm going to have to do each of the following:

100×20
100×5
40×20
40×5
3×20
3×5

– and then add them all together.

Long Multiplication

The first multiplication method, and the one most people will have been shown at school, is called long multiplication. You set out the sum in the usual way and then start by multiplying the units digit of the bottom number by each digit in the top number:

$$\begin{array}{cccc} & 1 & 4 & 3 \\ \times & & 2 & 5 \\ \hline \end{array}$$

3×5=15, so I put the 5 into the units column and carry the 1 to the top of the tens column:

$$\begin{array}{cccc} & & 1 & \\ & 1 & 4 & 3 \\ \times & & 2 & 5 \\ \hline & & & 5 \end{array}$$

Then I do 4×5=20 and, adding the carried 1, get 21 as my total. Put the 1 into the tens column (since we have in fact just worked out 40×5 and the 21 actually represents 210) and carry the 2 into the hundreds column.

$$\begin{array}{cccc} & 2 & 1 & \\ & 1 & 4 & 3 \\ \times & & 2 & 5 \\ \hline & & 1 & 5 \end{array}$$

Now I do 1×5=5 and add the carried 2 to get 7, which I write in the hundreds column because I'm actually doing 100×5 and the carried 2 is worth 200. In these three easy sums, I've already done half the calculations on the list.

$$\begin{array}{cccc} & 2 & 1 & \\ & 1 & 4 & 3 \\ \times & & 2 & 5 \\ \hline & 7 & 1 & 5 \end{array}$$

Now I repeat what I've done but using the 2 in 25, writing the answers on a new row underneath what I've done so far.

It's really important to remember that the 2 represents 20. My first calculation will be 2×3=6, but I mustn't put my answer in the units column, because the sum we're working out is in fact 20×3=60. When I do long multiplication, I like to put a 0 in the units column so that I don't accidentally write any answers there.* It also helps to rub or cross out the carries from the first set of calculations, so that they don't get confusing:

$$\begin{array}{cccc} & 1 & 4 & 3 \\ \times & & 2 & 5 \\ \hline & 7 & 1 & 5 \\ & & & 0 \end{array}$$

* If the bottom number of the multiplication had a third digit, something in the hundreds column, you'd have to put two 0s in on your third row of multiplying.

To cut a long story short, we work out 2×3=6, 2×4=8 and 2×1=2, and write those answers down:

$$\begin{array}{r} 143 \\ \times\ 25 \\ \hline 715 \\ 2860 \end{array}$$

In a few easy steps, I've done all the calculations on my list, albeit in easier forms. All that remains now is to add 715 and 2860, which is easy enough as they're set out in a sum already. So 143×25=3575. We have successfully broken down a multiplication that would be very tricky to do mentally into a series of much easier bits of arithmetic.

The Grid Method

The second way of doing multiplication is the grid method, which is basically a graphical way of approaching the problem. It works on the premise that, any time you multiply two numbers, you are in fact working out the area (see p.104) of a rectangle whose sides are the same length as those two numbers.

So for our calculation of 143×25, we could make a rectangle representing those numbers, and then split 25 into the more manageable chunks of 20 and 5, and 143 into 100, 40 and 3, to turn my rectangle into six smaller rectangles, each with an area equal to its sides multiplied together:

	100	40	3
20	20x100=2000	20x40=800	20x3=60
5	5x100=500	5x40=200	5x3=15

You'll notice, and this is the beauty of maths, that the areas of these six rectangles correspond exactly with the six calculations I identified on p.21. They look a bit tricky at first, but on second glance we can see that they mainly involve 0s, which I can ignore and add on to my answer later (i.e. 20×100 is 2×1 with three 0s on the end).

So the area of the large rectangle is equal to the areas of the small rectangles added together: 2000+800+60+500+200+15=3575. Once again, six relatively easy multiplications and a spot of addition have given us our answer.

Gelosia

Gelosia is a multiplication method that some people love and others just hate. It was first used in ancient India (hence its other name: Hindu Grid), while the name gelosia comes from an Italian word for a kind of lattice, which will make sense once we get started.

That said, getting started is a rather more laborious process with gelosia. Rather like the grid method, I set my numbers up on each side of a rectangle, but my rectangle is divided into squares, each of which is halved diagonally:

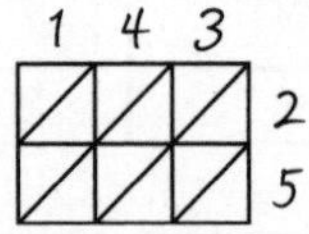

In each box, I write the answer I get from multiplying the number at the head of the column with the number at the end of the row. You can do them in any order you like, so let's start at the bottom right. I have 3 at the top and 5 at the side, so I do 3×5=15 and write my answer in the box, like so:

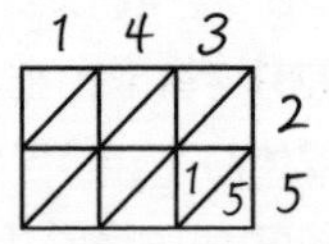

I then fill in the rest of the grid. If the multiplication comes to less than 10, you write 0 in the top half of the square.

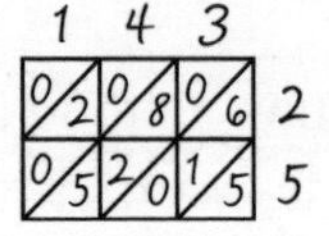

So far so good. But the really clever bit of gelosia is how we now derive an actual answer. Imagine that the grid is in fact a load of diagonal stripes:

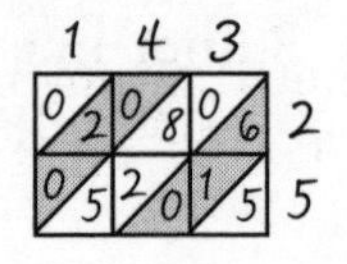

So we now have two grey stripes and three white stripes. Add up the numbers in each stripe and write the totals along the bottom, at the end of each stripe. Because carrying is likely to get involved, do the addition from right to left. The stripe on the right is simply 5 – that was easy – and the next one along is 0+1+6=7.

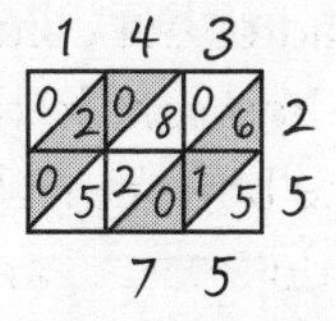

Now we hit a problem, since 5+2+8+0=15, but all we need to do in gelosia is carry the 1 into the next stripe along. So 1+0+2+0=3, and the final stripe is just 0:

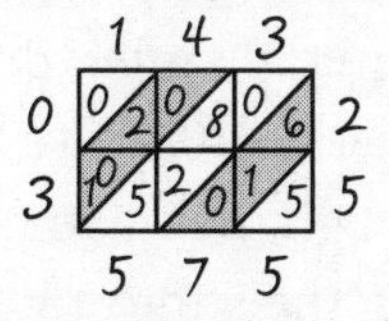

Our work is done. We get the answer by reading down the left-hand side and then across the bottom: 03575, AKA 3575.

Division

Division, as the name suggests, is all about dividing things up. It tends to inspire horror in anyone who's not very fond of maths, but the good news is that division can easily be solved by thinking about multiplication instead.

For example, the calculation 20÷5 asks the question, 'What is 20 when split into 5 equal parts?', or indeed, 'What number multiplied by 5 gives 20?' We know that 4×5=20, so 20÷5=4. But what about when things get a bit harder?

The method known as short division works well for when you're dividing by 12 or less, since we know our times tables up to 12×12. We'd set up 156÷12 like this:

$$12\overline{)\,1\quad 5\quad 6}$$

We take each digit of the bigger number in turn, starting from the left, and ask ourselves how many times the smaller number goes into it. So the first question I would ask myself is, 'How many times does 12 go into 1?' Answer: 0 times, so I write a 0 above the 1. We can't just forget about the 1, though, so we remainder it and write it next to the 5, turning the 5 into 15:

$$\begin{array}{r} 0\quad\phantom{{}^{1}5}\quad \\ 12\overline{)\,1\quad {}^{1}5\quad 6} \end{array}$$

How many times does 12 go into 15? Answer: once, with a remainder of 3. I write the 1 above the 15 and the remainder 3 next to the 6, turning it into 36:

$$\begin{array}{r} 0\quad 1\quad\phantom{{}^{3}6} \\ 12\overline{)\,1\quad {}^{1}5\quad {}^{3}6} \end{array}$$

How many times does 12 go into 36? Answer: 3 times exactly. Job done! 156÷12=13.

$$\begin{array}{r|l} & 0\ \ 1\ \ 3 \\ \hline 12 & 1\ \ {}^{1}5\ \ {}^{3}6 \end{array}$$

Here's a similar example, but with a twist: 102÷4. I start off in the same way, but there's something fishy when I get to this stage:

$$\begin{array}{r|l} & 0\ \ 2\ \ 5 \\ \hline 4 & 1\ \ {}^{1}0\ \ {}^{2}2 \end{array}$$

Four goes into 22 five times, but I have a remainder of 2. Where on earth do I put it?

Decimals (see p.54) come in handy here, and by remembering that 102 is the same as 102.0 and adding a decimal point after 102, I can keep adding 0s to the number without changing its value. The 0s allow me to use my remainder and continue working out the sum, giving:

$$\begin{array}{r|l} & 0\ \ 2\ \ 5.\ 5 \\ \hline 4 & 1\ \ {}^{1}0\ \ {}^{2}2.\ {}^{2}0 \end{array}$$

So 102÷4=25.5.

Long Division

There are many highly talented mathematicians who

quake at the thought of doing long division. The truth of the matter is that you can get by with short division and never have to worry about it, but when you divide by larger numbers the remainders get larger and larger and become much harder to squeeze in, as well as being incredibly difficult to work out in your head. Long division helps you set it all out neatly and lets you do a subtraction sum to work out the remainder.

Let's do a typically horrific example: 9594÷78. I set up in the same way as before:

```
   _________
78| 9 5 9 4
```

Technically, I should start off with 78 into 9, but I already know it won't go. I should write a 0 above the 9 and then a little 9 next to the 5, but I still end up doing 78 into 95, which is written there already. Right, so 78 goes into 95 once. I write in the 1 above the 5, but instead of working out the remainder in my head, I set up a simple subtraction:

```
      1
   _________
78| 9 5 9 4
  - 7 8
   ----
```

I don't need the 9 or the 5 anymore, so I can borrow in order to work out the subtraction:

```
        1
   ___________
78| 8̶9 ¹5 9 4
  -  7  8
    -----
     1  7
```

That sum tells me that the remainder of 95÷78 is 17. In short division, I would have had to do this in my head and then try to write 17 next to the second 9 of 9594. In long division, I bring the second 9 down to the remainder, like this:

```
          1
   78 | 8 9  1 5  9  4
    -     7    8  ↓
        ----------
          1    7  9
```

Now I ask myself how many 78s go into 179. Well, two 78s are 156, so three 78s will be too much. I put a 2 into my answer at the top, write 156 under 179, and do another subtraction to work out that the remainder of 179÷78 is 23:

```
          1    2
   78 | 8 9  1 5  9  4
    -     7    8  ↓
        ----------
          1    7  9
    -     1    5  6
        ------------
               2  3
```

I then bring down the last digit of 9594 and write it at the end of my remainder:

```
          1    2
   78 | 8 9  1 5  9  4
    -     7    8  ↓  ↓
        ----------
          1    7  9  ↓
    -     1    5  6  ↓
               2  3  4
```

As good luck would have it, 78 goes into 234 exactly 3 times, so my remainder calculation leaves me with 0:

```
        1  2  3
78 | 8̶9 ¹5  9  4
  -  7  8  ↓  ↓
     1  7  9  ↓
  -  1  5  6  ↓
        2  3  4
  -     2  3  4
              0
```

Ta-dah! 9594÷78=123.

Long division seems trickier than it actually is because it can be hard to remember what to do next, but you are in fact not doing anything that you don't do mentally in short division.

COMPLICATIONS IN ARITHMETIC

Arithmetic is, on the whole, fairly straightforward, but there are a number of fiddly annoyances that conspire to complicate matters.

INDICES

Indices are a sort of shorthand code for when we multiply numbers by themselves. If I want to multiply five 2s together, for instance, I can write it like this:

$$2 \times 2 \times 2 \times 2 \times 2 = 2^5$$

The small 5 next to the 2 is an index (plural indices), AKA a power or an exponent, and it means that the 2 is multiplied by itself five times. This way of writing it is nice and neat, but it's very important not to confuse 2^5 (32) with 2×5 (10).

There are several ways to say 2^5 out loud. The most formal would be '2 to the fifth power', but you're more likely to hear '2 to the power of 5' or the snappier '2 to the 5'.

A couple of indices have special names, and you've probably used them before without realizing it. A number with an index of 2 means that the number is multiplied by itself once, and we usually say 'squared' rather than 'to the power of 2':

$$\text{'4 squared'} = 4^2 = 4 \times 4 = 16$$

It's called 'squared' because you multiply a number by itself in order to find the area of a square.

The other index with a special name is 'cubed', and this denotes anything with an index of 3:

$$\text{'4 cubed'} = 4^3 = 4 \times 4 \times 4 = 64$$

It's called 'cubed' because it's the calculation you need to do in order to find the volume of a cube. If you simply can't wait to start calculating the volume of cubes, turn to p.176.

Square Roots

It is possibile to 'unsquare' a number, too – it's called finding the square root and the symbol looks a bit like a tick:

$$\sqrt{25}$$

This is asking the question: 'What number multiplied by itself gives 25?' and the answer, of course, is 5. So $5^2=25$ and $\sqrt{25}=5$.

You can have cube roots too:

$$\sqrt[3]{8}$$

'What number multiplied by itself three times gives 8?' 2×2×2=8, so the cube root of 8 is 2.

Order of Operations

When you need to do more than one operation – that's a sum, in plain English – how do you know which to do first?

Take 4×5+6. If I did the multiplying first, I would get:

$$4\times5+6=20+6=26$$

But if I did the addition first, I would get:

$$4\times5+6=4\times11=44$$

Each approach gives a wildly different answer, and mathematicians are generally not amused by simple arithmetic having two wildly different answers. To make things clearer, they have invented an order in which to carry out operations, which you can remember using the mnemonic BIDMAS:

Brackets
Indices
Division
Multiplication
Addition
Subtraction

Anything inside a bracket gets done first and any subtraction comes last.

So 4×5+6 is 20+6 rather than 4×11, and our answer is 26 rather than 44. If you wanted the sum to mean 4×11 rather than 20+6, you'd need to put 5+6 in brackets to ensure it gets done first:

$$4\times(5+6)=44$$

Negative Numbers

Negative numbers are a very good example of 'pure' as opposed to 'applied' maths, by which I mean they are theoretical rather than tangible. I cannot hold -3 apples.

When I spend too much money, the bank does not issue me with a negative tenner.

In the real world, we mainly encounter negative numbers in terms of debts and temperatures, neither of which are particularly nice to think about. In the more nebulous world of maths, however, we need to be able to do everything that we can do with positive numbers with negative numbers, too.

The following sums involving positive and negative numbers are all correct, but some of them take a bit of effort to get your head around:

$$5+8=13$$
$$-5+8=3$$
$$-5-8=-13$$
$$5-8=-3$$
$$5--8=13$$

Many people find a number line helpful when dealing with negative numbers. This is a bit like a thermometer lying on its side, and can be as long or short as we need it to be:

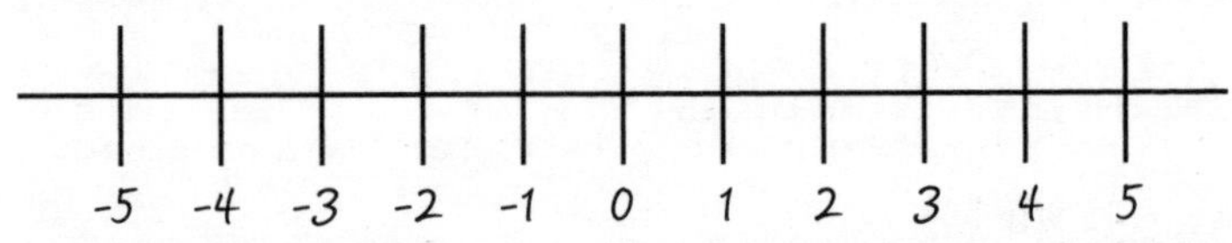

Now imagine the first number as your starting position, and then use the second number to plot your route to an answer. If you imagine yourself standing on the line with your face towards the higher numbers and your back to the lower numbers, this will be all the more simple.

When your sum asks you to add something, keep facing the higher numbers; when it asks you to subtract something, turn around to face the lower numbers. If the number you need to add or subtract is a positive number, walk forward in whichever direction you happen to be facing; if the number is negative, walk backwards. Easy!

Let's use these rules to work out the five previous sums.

- To work out 5+8, start at 5 and take 8 paces forward to reach 13.
- To work out -5+8, start at -5 and take 8 paces forward to reach 3.
- To work out -5–8, start at -5, turn around (since you're subtracting) to face the lower end of the scale, and take 8 paces forward to reach -13.
- To work out 5–8, start at 5, turn around and take 8 paces forward to reach -3.
- To work out 5--8 (eek!), simply start at 5, turn around to face the negative numbers, and then walk backwards (since it's a negative number) to reach 13.

'TWO MINUSES MAKE A PLUS'

It's worth pausing for breath at this stage since that last sum was rather a taxing one. Double negatives can be a real nightmare to deal with, and you were probably taught at school that 'two minuses make a plus' – but *why*? How on earth can 5−−8 be the same as 5+8?

If you do a simple subtraction – let's say 10−2 – the value of your original number reduces; in this instance, you are left with 8. If you did 10−1, your original number would still be reduced but by a smaller margin, and a calculation of 10−0 would leave your original number unchanged. It therefore follows that 10−−1=11, since it has to go in the opposite direction to 10−1. Subtracting a negative number creates a negative reduction, and a negative reduction is positive.

10−2=8 (reduction of 2)
10−1=9 (reduction of 1)
10−0=10 (reduction of 0)
10−−1=11 (reduction of -1, i.e. an increase of 1)

This all looks more complicated than it really is: we can't

have 10–2 and 10--2 giving the same result, so subtracting a negative takes you in the opposite direction to normal subtraction.

Commutativity

Commutativity is a fancy way of saying that it doesn't matter in which order you do your sum, as long as all the positives and negatives stay with the right numbers. The sum -5+8 can be broken down into a -5 and a +8 and then rearranged into +8–5, which is much easier on the eye.*
Most importantly, the answer stays the same whichever way you write it:

$$-5+8=+8-5=3$$

Multiplying and Dividing Negatives

When it comes to multiplying and dividing negative numbers, you need to keep your wits about you. Tricksy fellows, these negatives.

* Obviously we don't tend to bother writing a + before every positive number – I'm just, as maths teachers like to say, 'showing my workings'.

For multiplying, I find that thinking in terms of 'lots of' really helps. Consider these:

3×4	means 3 lots of 4	$=4+4+4$	$=12$
3×-4	means 3 lots of -4	$=-4+-4+-4$	$=-12$
-3×4	means 4 lots of -3	$=-3+-3+-3+-3$	$=-12$

Because multiplication works either way round (good old commutativity), I can think of -3×4 as either 4×-3 or 4 lots of -3. The tricky bit comes when you multiply two negatives. To explain this, I'm going to branch out into division for a minute.

I hope you'll agree that 10÷2=5. The sum is asking, 'What is half of 10?', to which the answer is 5. But if the sum were -10÷2, the question would become, 'What is half of -10?' The answer to this is -5, because two lots of -5 make -10.

$$-10\div2=-5$$

This is the same as asking, 'How many lots of 2 make -10?', and it follows that -5 lots of 2 make -10:

$$-5\times2=-10$$

Knowing what we do about the relationship between 2, 5 and 10, we can switch the negatives around to come up with a similar sum:

$$10\div-2=-5$$

This is the same as asking, 'How many lots of -2 make 10?', to which the answer is -5. If we juggle things around as we did with the previous sum, we end up with a slightly weird result:

$$-2 \times -5 = 10$$

Two negatives, when multiplied, give a positive.

There are two rules to bear in mind when multiplying and dividing numbers:

1. If both numbers are positive or negative, the answer is positive.
2. If one number is positive and one number is negative, the answer is negative.

MAKING THINGS EASIER

For a subject with so many absolutes and hard-and-fast rules, it is perhaps surprising that there are a handful of things designed to make maths nice and vague.

Rounding 'to the Nearest'

There are several ways to round, each with the aim of making the answer as precise as possible within defined parameters of accuracy. Let's take a really simple example. I want to round 6 to the nearest 10, so I have to work out whether 6 is closer to 0 (0×10) or 10 (1×10), and then round it up to 10. If I wanted to round 4 to the nearest 10, my answer would be 0.

'But what about 5?' I hear you cry. 'It's exactly halfway between the two!'

The convention is that we round up when the number is exactly in the middle of two options. There's a gentle optimism involved in that, which I quite like.

We can round to any number we like. If 35,729 people came to a football match, I could round this in the following ways:

35,729 =35,730 (to the nearest 10)
=35,700 (to the nearest 100)
=36,000 (to the nearest 1000)
=40,000 (to the nearest 10,000)

Estimating

Estimating, contrary to what its name suggests, is very different to guessing. To estimate something in maths, you do a calculation involving rounded numbers. Your answer will not be exactly correct, but it will give you a rough idea of what you'd get if you did the full calculation.

For example, if I were asked to work out 693×31, it would be useful for me to have an estimate of the answer before I did the long multiplication. That way, if my answer is wildly different, I know that I've probably made a mistake somewhere and need to check my working.

A decent estimate would be to say that 693 is 700 (to the nearest hundred) and 31 is 30 (to the nearest 10). Since 7×3=21, I can quickly work out that 700×30=21,000. When I do my long multiplication and declare that 693×31=21,483, I can be fairly sure from my 'guesswork' that I haven't made a mistake.

FRACTIONS

We are all used to what mathematicians call integers, which is just a fancy word for whole or counting numbers such as 1, 12 or 473. It all gets a bit more fiddly when we start dealing with the numbers between these numbers: fractions. Fractions are, unfortunately, vital to maths because of the infinite number of fractions between any two whole numbers. There are even an infinite number of fractions between any two fractions.

But why do fractions matter? Well, in the real world, fractions help us divide things up into portions: you might give someone half of your chocolate bar, or break it into fifths to share between five friends. Of course, in the real world, there is a limit to the number of people you can share one chocolate bar with, whereas in maths, you could keep making fractions of fractions of fractions and never run out of numbers.

VULGAR FRACTIONS

When we talk about fractions, we are more often than not referring to vulgar fractions, which are expressed as one number above another. The value of the fraction is equal to the top number, the numerator, divided by the bottom number, the denominator:*

$$13/25$$

There are two ways to think about 13 twenty-fifths. Sticking with the chocolate-bar analogy, imagine your bar had been divided into 25 equal parts – that is, 25 twenty-fifths. If you had 13 of those parts, you would have that fraction of the bar.

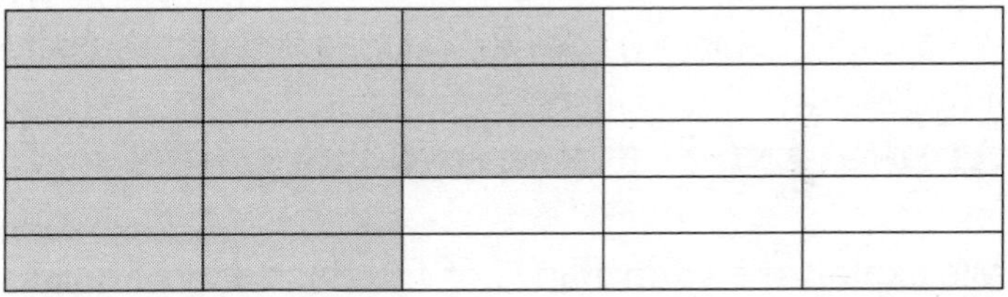

The other way is to think about it purely in terms of numbers. You have 13 and divide it by 25. Since 13 is smaller than 25, your answer will be smaller than 1: a fraction.

Much of the arithmetic of fractions relies on understanding a very important concept called the equivalence of fractions, which sounds considerably more complicated than it is. When we write down whole numbers – 247, for instance – they each have a unique value. There is no way that I can write the number 247

* The sign we use for division (÷) is, in fact, a little diagram of a fraction, with dots to show where numbers should go.

using different digits. With fractions, you'll no doubt be pleased to hear, there are infinite ways of expressing the same value.

As we learn from a very young age – mainly from being taught how to tell the time – half of something is the same as two quarters. And there are myriad ways of expressing the same amount:

$$1/2=2/4=3/6=4/8=50/100=85/170=\frac{123456}{246912}$$

Adding and Subtracting Fractions

We exploit the equivalence of fractions when we add or subtract fractions, since we can only add or subtract fractions with the same denominator. We can add halves together, or fifths, but not halves *and* fifths. If we absolutely have to add halves and fifths together, we need to find the lowest common denominator – that is, the lowest number that both halves and fifths can be transformed into. Here's an example:

$$1/5+1/2$$

A commonly perpetrated schoolboy error is to add the tops and the bottoms to get two sevenths. But two sevenths is less than the half we started with, so is clearly not the correct result. To find the lowest common denominator, we need to work out the lowest number that both 5 and 2 go into – in this case, 10 – and then turn $1/5$ and $1/2$ into tenths.*

The golden rule for equivalence is to do the same thing to the top and the bottom. To turn fifths into tenths, we multiply the denominator by two (5×2=10), so we have to do the same to the numerator:

$$1/5 = 2/10$$

Another way of thinking about this goes back to chocolate bars – if I have a bar that comes in five chunks and cut them all in half to get tenths, two tenths of the bar is the same as one of the original chunks.

I can do a similar thing with the halves, so my original sum becomes:

$$1/5 + 1/2$$
$$= 2/10 + 5/10$$
$$= 7/10$$

Subtraction works in a similar way to addition: find a common denominator and do a regular subtraction sum.

---

* It's often the case that the lowest common denominator is the product of the two denominators (in this instance, 5×2=10), although with larger numbers, you can usually cancel down (see p.51) in order to find an even lower lowest common denominator. While 6×8=48, for example, 6 and 8 both also go into 24.

Multiplying and Dividing Fractions

Good grief, why would anyone want to do such a thing?

When it comes to multiplying and dividing, it's important to wrap your head around how closely related they are:

Dividing by 4 is the same as
multiplying by a quarter –
$3 \div 4 = 3 \times 1/4$

– and you don't need to be a rocket scientist to realize that three lots of a quarter is three-quarters.

Dividing a number by 2 is the same as finding half of it. This is where we get the oft-repeated maths-teacher mantra, 'OF MEANS TIMES!'

$$8 \div 2 = 8 \times 1/2$$

This is all very well for easy examples such as these, but what about for more daunting fractions? The good news is that multiplying fractions is one of the easiest things to do in maths – easier than adding them – because we simply multiply the numerators and the denominators:

$$\begin{aligned} &3/4 \times 5/7 \\ &= \frac{3 \times 5}{4 \times 7} \\ &= 15/28 \end{aligned}$$

The sum above asks the question, 'What is three-quarters of five-sevenths?', so let's go back to our much-handled chocolate bar to visualize how the calculation works. Imagine the chocolate bar has seven chunks in it, and we are specifically thinking about five of them:

As I'm dealing with quarters and sevenths, I need to find the lowest common denominator to translate between the two. The lowest common denominator of 4 and 7 is 28, so my 5 sevenths become 20 twenty-eighths:

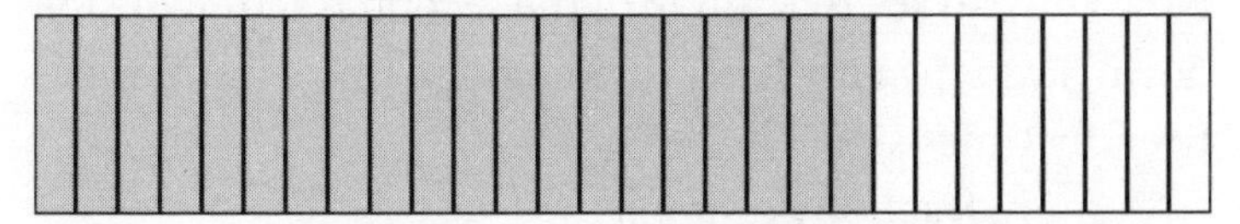

Chuck away (or, more likely, eat) the unwanted $2/7$, and split the remaining $5/7$ into quarters. As there are 20 pieces, each quarter is worth 5 pieces:

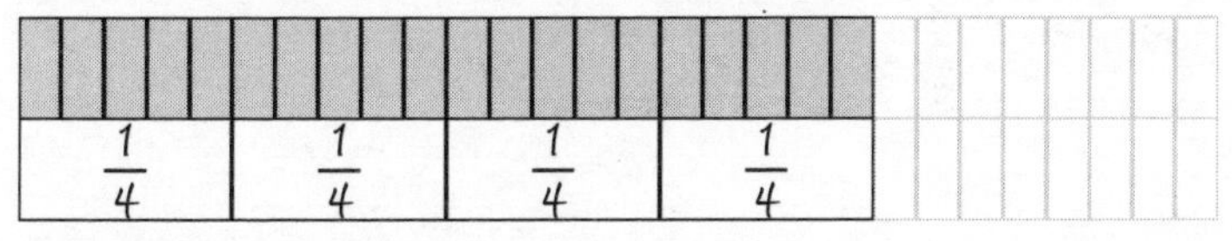

Counting up how many of the pieces make up three-quarters of what's left of the bar, we can see that $3/4$ of $5/7$ is $15/28$.

Dividing something by, say, 5, is the same thing as finding one fifth of it. Or dividing by 17 is the same as finding one seventeenth of your number. Of means times! Taking a straightforward example:

$$6 \div 2 = 6 \times 1/2$$

– and both equal 3.

To really see what's happened here, let's turn the whole thing into fractions. To turn whole numbers into fractions, we just put them over 1, because if you divide something by 1, you get what you started with.

$$6/1 \div 2/1 = 6/1 \times 1/2$$

Now you can see that, if I turn the second fraction upside down, I can turn my nasty dividing sign into a much more friendly times sign.

A more fractiony example:

$$3/5 \div 5/6$$

Change ÷ to × and turn the second fraction upside down:

$$= 3/5 \times 6/5$$
$$= 18/25$$

CANCELLING DOWN

Because multiplying fractions simply involves multiplying the numerators and then the denominators, it doesn't actually matter which numerator is on top of which denominator: the answer will be the same. Look at this sum:

$$10/13 \times 7/15$$

While I can work out 10×7 in my head, 13×15 is a bit more of a thinker. But if I swap the denominators –

$$10/15 \times 7/13$$

– at least one of the fractions becomes easier to deal with, since $^{10}/_{15}$ can easily be cancelled down into $^{2}/_{3}$.

Cancelling a fraction down means reducing it to its lowest possible form. The smallest number (other than 1) that goes into both 10 and 15 is 5, so I can divide both 10 and 15 by 5 to turn $^{10}/_{15}$ into $^{2}/_{3}$, a fraction of exactly the same value.

Now we only need to multiply 13 by 3 rather than by 15:

$$10/15 \times 7/13$$
$$= 2/3 \times 7/13$$
$$= 14/39$$

Mixed Numbers

When we're adding and subtracting and generally mucking about with fractions, whole numbers can become involved with fractions to give what are known as mixed numbers. Let's take this sum as an example:

$$5/6+3/8$$

The lowest common denominator here is going to be 24, as it's the lowest number both 6 and 8 go into, so we'll turn both fractions into twenty-fourths:

$$=\frac{5\times4}{6\times4}+\frac{3\times3}{8\times3}$$
$$=20/24+9/24$$
$$=29/24$$

Oh dear, our answer is top-heavy: it has a numerator that is bigger than the denominator. We know that $24/24=1$, so a top-heavy answer indicates a number that is greater than 1.

$$29/24=24/24+5/24$$
$$=1+5/24$$
$$=1\,5/24$$

Our answer, $1\,5/24$, is a mixed number as it mixes a whole number and a fraction.

We can add and subtract mixed numbers, dealing with the whole numbers and the fractions separately.

Multiplying Mixed Numbers

Although multiplying fractions is surprisingly easy, multiplying mixed numbers adds a touch of spice to proceedings. If you deal with the whole numbers separately, you'll come up with the wrong answer, so we need to turn our mixed numbers back into top-heavy fractions. How is this done? Take a look at this:

$$3\tfrac{3}{4} = \tfrac{12}{4} + \tfrac{3}{4}$$
$$= \tfrac{15}{4}$$

To turn the 3 into a fraction, I had to work out how many quarters make a whole – 4 – and then multiply that by 3 to find that 3 wholes = 12 quarters.

Now to use that fraction in a multiplication sum:

$$3\tfrac{3}{4} \times 3\tfrac{1}{9}$$
$$= \tfrac{15}{4} \times \tfrac{28}{9}$$

At this stage, I'll swap the denominators to make life easier for myself –

$$= \tfrac{15}{9} \times \tfrac{28}{4}$$

– and this is easily cancelled down:

$$= \tfrac{5}{3} \times \tfrac{7}{1}$$
$$= \tfrac{35}{3}$$
$$= 11\tfrac{2}{3}$$

Dividing Mixed Numbers

Dividing mixed numbers is similar to multiplying them in that you need to create top-heavy fractions before you begin:

$$4\frac{1}{2} \div 2\frac{2}{5}$$
$$= \frac{9}{2} \div \frac{12}{5}$$

As with regular division, we then change the ÷ into × and turn the second fraction on its head –

$$= \frac{9}{2} \times \frac{5}{12}$$

– before swapping denominators –

$$= \frac{9}{12} \times \frac{5}{2}$$

– and cancelling down:

$$= \frac{3}{4} \times \frac{5}{2}$$
$$= \frac{15}{8}$$
$$= 1\frac{7}{8}$$

If you can follow all of that, you can handle pretty much anything a fraction can throw at you!

DECIMALS

Decimals, numbers written to the right of the decimal point, work in exactly the same way as ordinary numbers. Columns and place value still reign supreme, so we can add and subtract decimals in the normal way.

OK, there's one big difference, and that's that you need to imagine the decimal point is a sort of mirror: with decimals, each column is ten times bigger than the one to the right of it, or ten times smaller than the one to the left of it.

It's worth remembering that any decimal can have any number of 0s on the right hand side of it without changing the value of the number. So 123.456 is the same as 123.4560, which is the same as 123.456000000000.

Now for a bit of an eye-opener. Between any two whole numbers, there is an infinite number of other decimal numbers. And lurking in the background between any two decimals, no matter how close together they are, there is still an infinite number of other decimals.

Adding and Subtracting Decimals

The next thing we need to be able to do is arithmetic with decimals. You'll no doubt be relieved to find that adding and subtracting decimals is exactly the same as with whole numbers. In fact, I reckon it's even a bit easier. As long as you line up the decimal points, it's pretty hard to mess up.

$$\begin{array}{r} 27.34 \\ +\ 11.25 \\ \hline 38.59 \end{array}$$

The one thing to watch out for is subtracting decimals that have a different number of columns:

$$\begin{array}{r} 2.26 \\ -\ 3.438 \\ \hline \end{array}$$

It's easy and tempting to assume there's nothing to do to the 8 and to whack down an 8 at the end of the answer. But if you bear in mind that every decimal number has an infinite number of 0s at the end of it, your sum will look a little different:

$$\begin{array}{r} 5.260 \\ -\ 3.438 \\ \hline \end{array}$$

Now that you need to borrow from a 0 to subtract the 8, your answer will look like this:

$$\begin{array}{rcccc} & 4 & & & 5 \\ & \not{5} & .\ {}^{1}2 & \not{6} & {}^{1}0 \\ - & 3 & .\ 4 & 3 & 8 \\ \hline & 1 & .\ 8 & 2 & 2 \end{array}$$

MULTIPLYING DECIMALS

When you multiply by a number that is less than one, the answer gets smaller. If I try to work out 0.4×10, I am asking, 'What is 0.4 lots of 10?' Since 0.4 is less than 1, I'll end up with less than one lot of 10. But how do I work out the answer?

The simplest technique I've come across is to pretend

that there are no decimal points in the question. The 0.4×10 becomes 04×10. This is the same as 4×10, which is 40. All I need to do now is put the decimal point back in so that the answer has the same number of decimal places as the question. I bumped my 0.4 up by one decimal place to turn it into 4, so I just need to bump my answer – 40 – back down again in order to find the real answer: 4.0, AKA 4.

Here's another example: 12.4×3.2. I can estimate (see p.42) that my answer will be something a bit more than 36, because 12×3=36.

First things first: pretend there are no decimal points. 12.4×3.2 becomes 124×32, which is a long multiplication:

$$\begin{array}{cccc} & & 1 & \\ & 1 & 2 & 4 \\ \times & & 3 & 2 \\ \hline & 2 & 4 & 8 \\ 3 & 7 & 2 & 0 \\ \hline 3 & 9 & 6 & 8 \end{array}$$

So 124×32=3968. Now I have to put the decimal point back in the right place. There are two decimal places in the question (12.$\underline{4}$×3.$\underline{2}$), so I need to have two in my answer: 39.$\underline{68}$. This answer seems reasonable, given that I was expecting something a bit higher than 36.

If I multiply two decimals that are less than 1, my answers get very small very quickly. For instance, 0.2×0.1=0.02 (it's the same as 2×1 with 2 decimal places added in).

Dividing Decimals

There are two situations you can find yourself in when dividing decimals. One is good and one is slightly evil.

Good Decimal Division

This is where the first number in the division is a decimal, but the second one isn't. A good example is 17.56÷4, and I can proceed in the normal way because the number I am dividing by is a whole number:

$$\begin{array}{r} 4\,.\,3\,9 \\ 4\,\overline{)\,1\,7\,.\,{}^{1}5\,{}^{3}6} \end{array}$$

The only thing I had to watch out for there was to make sure I remembered to put a decimal point on the top line as I did the question.

Evil Decimal Division

This is where the number you are dividing by is a decimal, e.g. 53.7÷0.2. Here I can't proceed as normal because I'd start off trying to divide 0.2 into 5 (25) and would soon run out of room for my enormous answer.

Instead, we have to exploit our old friend the equivalence of fractions (see p.45). Division sums are, after all, just a different way of writing fractions. I could write this question as:

$$\frac{53\cdot7}{0\cdot2}$$

This is a weird-looking fraction but if we can turn the denominator into a whole number, it'll be much easier to work with.

Remember that the equivalence of fractions dictates that the value of a fraction will remain the same as long as I do the same thing to the top and the bottom. If I multiply the top and bottom numbers by 10 – by moving the decimal point one space to the right – the fraction will look like this:

$$\frac{53\cdot7 \quad \times 10}{0\cdot2 \quad \times 10} = \frac{537}{2}$$

Now I can try my division again with these much larger numbers, safe in the knowledge that my new sum has the same value as my original sum, but is much easier to deal with.

$$\begin{array}{r|l} & 2\ \ 6\ \ 8\ \cdot\ 5 \\ \hline 2 & 5\ \ {}^{1}3\ \ {}^{1}7\ \cdot\ {}^{1}0 \end{array}$$

So 53.7÷0.2=537÷2=268.5.

It's only the number you're dividing by that needs to be whole. If I were working out 1.735÷0.012, I would multiply both by 1000 to give me 1735÷12, and my answer would be the same as the answer to the original sum. Evil division isn't so evil after all!

Just as multiplying a decimal by a power of 10 causes the decimal point to travel to the right, dividing by a power of 10 makes it shift to the left:

$$0.2 \div 10 = 0.02$$
(move the decimal point one place to the left)

$$0.2 \div 100 = 0.002$$
(move the decimal point two places to the left)

$$0.2 \div 1000 = 0.0002$$
(move the decimal point three places to the left)

Recurring Decimals

Every now and then, you'll find that a division gets stuck in a loop. If I do 1÷3, I'll get 3 remainder 1 ad infinitum:

$$\begin{array}{r|l} & 0\,.\,3\;3\;3\;3\;\ldots \\ \hline 3 & 1\,.\,{}^{1}0\;{}^{1}0\;{}^{1}0\;{}^{1}0\;\ldots \end{array}$$

This is known as a recurring decimal, and the shorthand way to indicate such an irritating number is to put a little dot over the offending digit: $1 \div 3 = 0.\dot{3}$.

A recurring decimal can also be one in which a group of digits repeats. If I do 1÷7, I'll get:

$$\begin{array}{r|l} & 0\,.\,1\;4\;2\;8\;5\;7\;1\;4\;2\;8\;5\;7\;1\;\ldots \\ \hline 7 & 1\,.\,{}^{1}0\;{}^{3}0\;{}^{2}0\;{}^{6}0\;{}^{4}0\;{}^{5}0\;{}^{1}0\;{}^{3}0\;{}^{2}0\;{}^{6}0\;{}^{4}0\;{}^{5}0\;{}^{1}0\;\ldots \end{array}$$

Here, we get a little recurring loop of 142857. To indicate

this in shorthand, you put a dot on the first and last digits of the pattern: $1 \div 7 = 0.\dot{1}4285\dot{7}$

PERCENTAGES

A percentage is the same as a fraction of 100, and is shown with the symbol %.

We come across percentages in everyday life far more often than we do fractions and decimals, for the simple reason that it can be hard to picture the value of a fraction in your head. If an item you want to buy is reduced by a third in one shop but by seven-twentieths in another, you'd be pretty stumped as to which was cheapest. By showing their reductions as percentages, shops make it easier for us to compare prices. One third is 33% (rounded to the nearest percent) whereas seven-twentieths is 35%.

We're also used to seeing percentages used in terms of interest, inflation and tax rates. Over the past few years, VAT has changed from 17.5% to 15% and back again, and it's far easier to appreciate what has happened when these are expressed as percentages. If the Chancellor suddenly announced, 'We're changing VAT from seven-fortieths to three-twentieths', there'd be quite a bit of head scratching and scribbling on backs of envelopes before anyone knew what on earth was going on.

If you need to do any arithmetic with percentages, the best thing to do is remember that a percentage is a fraction out of 100. If 99% of people dislike Brussels sprouts, it means that 99 people in every 100 people dislike Brussels sprouts.

It's also worth remembering that you are generally trying to work out a percentage *of* something – and, as we know only too well, of means times.

Imagine I am selling an item at auction and the deal is that the auctioneer gets 10% of whatever the item sells for. My item sells for £400, so I want to work out how much of that I owe the auctioneer.

$$\begin{array}{ccc} 10\% & \text{of} & £400 \\ \downarrow & \downarrow & \downarrow \\ \frac{10}{100} & \times & \frac{400}{1} \end{array}$$

A quick bit of arithmetic gives us an answer of $^{40}/_{1}$, i.e. £40.

CONVERTING FRACTIONS, DECIMALS AND PERCENTAGES

You probably already knew that 10% of 400 is 40 without going through all that rigmarole. This is because 10% is an easy percentage to deal with, because it has the same value as $^{1}/_{10}$.

There are a few values of fractions, decimals and percentages that most of us know off by heart, often without even realizing it. Here's a table of conversions that are handy to know:

Fraction	Decimal	Percentage
1/20	0·05	5
1/10	0·1	10
1/8	0·125	12·5
1/5	0·2	20
1/4	0·25	25
1/3	$0{\cdot}\dot{3}$	33·3
3/8	0·375	37·5
2/5	0·4	40
1/2	0·5	50
3/5	0·6	60
5/8	0·625	62·5
2/3	$0{\cdot}\dot{6}$	66·7
7/10	0·7	70
3/4	0·75	75
4/5	0·8	80
7/8	0·875	87·5
9/10	0·9	90
1	1	100

Sometimes you need to convert other fractions that aren't on this list. There are a few simple rules to follow for this.

If you look at the table, it's quite easy to see the link between decimals and fractions. The decimals are all between 0 and 1, and the percentages are between 0 and

100 – they are 100 times larger than the decimals. So to turn a decimal into a percentage, we multiply by 100 and stick a % on the end of it:

$$0.48 \rightarrow 0.48 \times 100 = 48\%$$

Now, fractions have the same value as their decimal counterparts (i.e. $1/4=0.25$), so if multiplying by 100 works for decimals, it should work for fractions, too. I'm going to pick $1/2$ to check this out on, because we know that we should end up with 50%:

$$1/2 \rightarrow 1/2 \times 100/1 = 100/2 = 50\%$$

It works.

And it follows that, if you multiply by 100 to turn fractions and decimals into percentages, you must have to divide percentages by 100 to get fractions and decimals. Let's have a go with 75%:

$$75\% \rightarrow 75 \div 100 = 0.75$$
$$75\% \rightarrow 75/100 = 3/4$$
(cancelled by dividing top and bottom by 25)

Success!

Now that we've turned both decimals and fractions into percentages, and vice versa, all that remains is for us to turn fractions and decimals into one another.

If you think about it, a fraction is basically a division: a numerator divided by a denominator. So if I want to know what that fraction is as a decimal, I just do that division. Let's check this with three-eighths, i.e. 3÷8:

$$8 \overline{\big)\, 3 \cdot {}^{3}0\ {}^{6}0\ {}^{4}0} \quad = 0 \cdot 3\ 7\ 5$$

As expected, dividing 3 by 8 gives us 0.375, which is three times more than 0.125, AKA ⅛. Hurray!

Turning decimals into fractions is slightly trickier. If I look at the digits in 0.375, I have 3 in the tenths column, 7 in the hundredths column and 5 in the thousandths column. This means that:

$$0.375={}^{3}/_{10}+{}^{7}/_{100}+{}^{5}/_{1000}$$

The lowest common denominator is 1000:

$$\begin{aligned}&{}^{3}/_{10}+{}^{7}/_{100}+{}^{5}/_{1000}\\ &=\frac{3\times100}{10\times100}+\frac{7\times10}{100\times10}+\frac{5}{1000}\\ &={}^{300}/_{1000}+{}^{70}/_{1000}+{}^{5}/_{1000}\\ &={}^{375}/_{1000}\end{aligned}$$

This all looks very complicated and we haven't even got an answer in its lowest terms yet. But what I want you to spot is that, if I take the digits of the decimal, I have that amount of whatever the final column is. So in the case of 0.375, I have 375 thousandths.

To finish off the calculation, we need to do a spot of

cancelling down, first by 25 –

$$375/1000=15/40$$

– and then by 5:

$$15/40=3/8$$

And once again we have 3/8. Got there in the end.

To end this chapter with a bit of fun, let's think about turning 0.85 into a fraction. The final column in 0.85 is the hundredths column, so I can announce with some confidence that 0.85 is 85 hundredths:

$$0.85=85/100$$
$$=17/20 \text{ (cancel top and bottom by 5)}$$

Now that the ins and outs of fractions, decimals and percentages are all you can think about, let's take a look at another topic that relies heavily on them.

PROBABILITY

When was the last time you went to the bookies and bet hundreds of pounds that your house would burn down at some point in the next year? What odds would they give you? How on earth would they calculate them?

When you pay your home insurance, this is effectively what you are doing. The insurance company will ask numerous questions about you and your property, work out the likelihood (or probability) of your house burning down based on values worked out by their clever mathematicians – and then charge you accordingly.

Probability often crops up in everyday life – from the more obvious examples such as playing the lottery or gambling, to the numerous times you take a chance on the perceived likelihood of something happening or not happening. If, for instance, you've been standing at the bus stop for 10 minutes and buses are due every 6 minutes, you might well deduce that a bus is probably going to show up very soon – just as soon as you give up and hail a taxi, in fact.

The field of probability deals with such events and their outcomes. An event is something that happens that may have a number of possible outcomes. Rolling a dice, for example, is an event with six possible outcomes.

SOME TECHNICAL TERMS, AND LOTS OF BORING DICE

Probabilities are usually expressed as fractions in any of their three forms (fractions, decimals and percentages), but first let's consider some everyday vocabulary as it applies to probability. For each, I have given a very dull, but accurate, example relating to dice.

Impossible Versus Possible

An event that is impossible has a probability of 0. It cannot happen.

It is actually quite hard to come up with an event that is truly impossible, as most things people come up with are certainly far-fetched but not genuinely impossible. Even school textbooks get it wrong, or rely on too many assumptions. 'It is impossible for me to go to the Moon tomorrow': granted, this is highly unlikely, but between now and tomorrow a friendly race of aliens could show up offering sightseeing tours of the dark side of the Moon. Of course, the chances of this happening may be astonishingly small, but they are not 0. You never know.

Anything that is not impossible is possible, so a mathematician will deem any event with a probability greater than 0 as possible. No matter how teeny-tiny the probability, if it is non-0, it is possible.

Boring dice example:
It is impossible to roll a 7 on an ordinary 6-sided dice.
It is possible to roll a 3 on an ordinary 6-sided dice.

Certain Versus Uncertain

Just as it can be very hard to come up with impossible outcomes, certain outcomes are tricky creatures, too.

While impossible outcomes have a probability of 0, certain outcomes have a probability of 1 (or 100%, if you're indicating probability with percentages). Textbooks are peppered with statements such as, 'It is certain the Sun will rise tomorrow', but, without wishing to be too nihilistic, there is an outside chance that something horrific might suddenly happen to either the Sun, the Earth, or both, which would prevent this from occurring – many science-fiction films rely heavily on this apocalyptic notion for their plot.

Uncertain outcomes can have any probability in the range between 0 and 1. Ironically, an outcome that is impossible is not uncertain, because you are certain that it has a probability of 0.

Boring dice example:
You are certain to roll a 1, 2, 3, 4, 5 or 6 on an ordinary 6-sided dice.

Evens

Smack in the middle of our spectrum of probability is the good old coin toss, the 50/50 gamble. Evens usually refers to an event with two possible outcomes that are equally (i.e. 50%) likely, such as tossing a coin and getting either heads or tails. You have an even chance of getting heads when you toss a coin.

Boring dice example:
You have an even chance of rolling a 1, 2 or 3 on an ordinary 6-sided dice.

Likely Versus Unlikely

Likely outcomes have greater than 50% chance of occurring – they are more likely (hence the name) to happen than not to happen, but they are not certain.

The words probably and probable are often used in reference to outcomes that are likely, while improbable refers to unlikely events – that is: events whose probability is greater than 0 but smaller than 50%.

Boring dice example:
It is likely that you will roll 5 or less on an ordinary 6-sided dice.
It is unlikely that you'll roll a 6 on an ordinary 6-sided dice.

WHY ALL THE DICE?

Most school textbooks use dice to explain probability, largely because they are commonplace in children's lives but also because, if we assume that the dice are completely fair or unbiased, all of the outcomes are equally likely, and a rule of probability can be exploited to invent doable maths questions.

The rule is that all the probabilities of all the outcomes of a particular event must add up to 1.

In the context of our boring dice, this is saying that:

The probability of rolling a 1
+ the probability of rolling a 2
+ the probability of rolling a 3
+ the probability of rolling a 4
+ the probability of rolling a 5
+ the probability of rolling a 6
= 1, i.e. it is certain that one
of these outcomes will occur.

If these outcomes are all equally likely, we can say that the probability of each of them occurring must be 1 divided by the number of outcomes, which in this case is 6. So the probability of rolling any particular number on a dice is $1/6$. Equally, if you had a fair 10-sided dice, the chance of rolling any particular number is $1/10$.

Once you've mastered all that, you can move on to rolling two dice and adding the totals, as you do in many board games. However, even though you have an equal chance of getting each number on each dice, your chances of getting a particular score when you add your two dice together are not equal. Why is this?

The possible outcomes of rolling two dice and adding the scores are anything between 2 (from a double 1) and 12 (from a double 6). But, just to confuse matters, there are multiple ways of getting some scores.

You were almost certainly forced to painstakingly draw this table in your maths exercise book at some point in your education. The **bold** numbers show my score on one dice and the <u>underlined</u> numbers are my score on the other. The table shows the totals I can get from the various outcomes.

+	**1**	**2**	**3**	**4**	**5**	**6**
<u>1</u>	2	3	4	5	6	7
<u>2</u>	3	4	5	6	7	8
<u>3</u>	4	5	6	7	8	9
<u>4</u>	5	6	7	8	9	10
<u>5</u>	6	7	8	9	10	11
<u>6</u>	7	8	9	10	11	12

There are 36 outcomes on this table and all of them are equally likely to occur when I roll two dice. It is just as likely that I will roll a 6 on one dice and a 6 on the other as it is that I'll roll a 3 on one and a 4 on the other.

But we know from playing Monopoly that 12 is an unlikely outcome. This is not because you are 'unlucky'; it is because the only way to get 12 is by rolling a double 6 – as the table shows, it's a $1/36$ chance. Rolling a double 1 is just as unlikely, but less exciting when you score it during a game of backgammon.

In contrast, there are six different ways of rolling a score of 7:

1 and 6, 2 and 5, 3 and 4, 4 and 3, 5 and 2, 6 and 1.

So the chance of getting a 7 is:

$$1/36 + 1/36 + 1/36 + 1/36 + 1/36 + 1/36 = 6/36 = 1/6$$

Whichever way you look at it, you are six times more likely to get a total of 7 than a double 6.

MATHEMATICAL VERSUS STATISTICAL PROBABILITY

In the realm of maths GCSE questions and card and dice games, the probabilities involved are mathematical. They are based on a finite number of possible outcomes. But in 'real life' examples such as horse races, football scores, insurance calculations and the like, the probabilities are calculated by gathering statistical information.

For instance, there is no mathematical formula for calculating the number of goals that a football team will score in a match. What you could do is look at the number of goals that team has scored in, say, the past five matches, and then work out how many goals, on average, they score each match. You could then use information to decide whether to bet with or against them.

Expected Outcomes

Whether a probability is mathematical or statistical, it is often useful to be able to predict certain things. If I throw a dice 100 times, how many 2s should I expect to get?

Well, the chance of rolling a 2 is $\frac{1}{6}$, so I should expect $\frac{1}{6}$ of my 100 rolls to be 2s. This works out at about 16.7, or 17 if we're rounding up (see p.41).

Unfortunately, were I actually to spend twenty minutes rolling a dice 100 times, I probably would not get sixteen or seventeen 2s, but this doesn't mean that my calculation is wrong. It simply means that, were I to have an awful lot of time on my hands and do my experiment over and over again, I could expect my average number of 2s per 100 rolls to end up being 16.7.

It is mathematically possible, however unlikely, for me to roll 100 2s in a row, or to get no 2s in my 100 rolls, in the same way as I might technically have won the lottery every week since it started, using the same six numbers. It's not impossible, but it's unlikely.

OR AND AND

Here are two questions that could be found in any school kid's maths book:

What is the probability of rolling a 5 or 6 on an ordinary dice?
What is the probability of rolling a 5 and then a 6 on an ordinary dice?

These two questions use mainly the same words, but the distinction between 'or' and 'and' leads to very different answers.

The way I think about it is this: OR implies having more options, more outcomes that satisfy the event. So in the case of an OR question, I add the probability of each separate outcome.

The chance of my rolling a 5 is $\frac{1}{6}$
The chance of my rolling a 6 is $\frac{1}{6}$
The chance of my rolling a 5 or 6 is, therefore,
$\frac{1}{6} + \frac{1}{6} = \frac{2}{6} = \frac{1}{3}$

A question with AND THEN adds to the list of things that have to happen and often the order in which they happen, too, reducing the chance of success:

The chance of my rolling a 5 is $1/6$
The chance of my rolling a 6 is $1/6$
The chance of my rolling a 5 and then a 6 is
$1/6 \times 1/6 = 1/36$

The Lottery

When you play the lottery, you select 6 numbers out of 49, and you need all of them to come up in the draw in order to win the jackpot. So imagine you are watching the draw with your ticket.

When the first ball comes, it could be any one of 49 numbers, 6 of which are yours. Your chance of getting a match is $6/49$.

The second ball comes. There are now only 48 possible outcomes, and, assuming the first ball was a match, you only have 5 numbers left. So your chance of getting this one is $5/48$.

The third ball comes. It has 47 outcomes and you have 4 numbers left: $4/47$.

I hope you're starting to see a pattern. This is a classic AND THEN-type problem, so we will be multiplying all these probabilities rather than adding them:

$$6/49 \times 5/48 \times 4/47 \times 3/46 \times 2/45 \times 1/44 = \frac{720}{10068347520}$$

This cancels down, roughly, to 1/13983816 – your chance of winning the jackpot is about 1 in 14 million. Based on what we know about expected outcomes, this implies that you'd have to play the lottery 14 million times in order to expect to win the jackpot once. If you played once a week, this would take you about 270,000 years.

Poker

Many probability questions centre on cards. If we assume that the deck is always well shuffled and complete (minus the jokers), your chance of being dealt any particular card is 1/52.

More experienced cards players are able to work out their chances of being dealt a particular card, or suit of card, and gamble accordingly. Good cards players are often said to 'make their own luck', but in reality they use a combination of practical skills: accurately remembering which cards have been dealt, knowing the likelihood of their getting a particular hand, and, in the case of poker, using the body language and habits of their opponents to guess what they might be holding.

Poker is a good example for us to look at, because the priority of the hands you can get is based on the probability of getting them. The probabilities are worked out from the number of ways in which you can end up with each type of hand, just as they are when rolling more than one dice.

If I start dealing cards, there are 52 outcomes for the first one, 51 for the second, 50 for the third, and so on. So on first glance there are going to be 52×51×50×49×48 potential five-card hands – almost 312 million different hands! But you have to take into account that lots of these hands will contain the same cards, just dealt in a different order. For example, if I am dealt the 3, 5, 7, 9 and 10 of clubs in that order, it's the same hand as if I got dealt the 10, 9, 7, 5 and 3 of clubs in that order. The order in which I receive the cards is not important; it's what type of hand they represent once they've all been dealt that matters.

It turns out that there are 120 different ways of ordering five cards, so I'll have to divide my total by 120 to find out how many different five-card hands there are:

$$(52\times51\times50\times49\times48)\div120=2{,}598{,}960$$

So there are the best part of 2.6 million different poker hands.

Now let's consider the best possible hand in poker: the Royal Flush. This is where you have the 10, Jack, Queen, King and Ace of the same suit in your hand. There are only four types of Royal Flush, one for each suit. So, of the 2.6 million hands you can get, only four of them can be a Royal Flush:

$$\frac{4}{2598960} = \frac{1}{649740}$$

Your odds of getting a Royal Flush are about 650,000 to 1. Compare this to the chances of getting one pair – I'll spare you the maths and just say that there are 1,098,240 ways of getting one pair, giving about a 42% chance.

Here's a list of the odds of each hand:

Hand	Number of combinations	Probability (%)
No Pair	1302540	50
One Pair	1098240	42
Two Pairs	123552	5
Three of a Kind	54912	2
Straight	10200	0.4
Flush	5108	0.2
Full House	3744	0.1
Four of a Kind	624	0.02
Straight Flush	36	0.001
Royal Flush	4	0.0002

The percentages here are quite rough, but you can see how quickly the odds drop after one pair. These odds don't take into account swapping cards in draws, but it does show why the poker hierarchy is as it is.

STATISTICS

Rarely is the word 'statistics' uttered without being prefaced by 'Lies, damned lies and…' The reason for this is that, while the figures themselves don't lie, the information we gather from them can easily be manipulated.

One of the aims of statistics is to give us information about a large number of things (a population) by looking at a small selection of those things (a sample). The human brain, though rather impressive, isn't great at looking through a long list of numbers and making deductions from the data. We rely instead on showing the information in a graph or chart and also on statistics themselves – numbers that tell us information about the data under consideration.

For instance, if I were a clothing designer instead of a maths teacher, I might want to know the 'average' height of a man in the UK. I could go and measure the heights of every man over the age of eighteen, but that would be incredibly dull. So I would need to choose a sample of men that would give me an accurate enough value.

But now for the tricky part: how do I choose my sample and make it fair? One way is to make the sample random: get a big list of every man in the country, get a computer to pick, say, 1000 at random, persuade or more likely pay them to let me measure their height, and hope that my 1000 men were representative of the whole of the country.

DATA

At this stage, I feel compelled to explain fully the word data, which is commonly misused. Data is the plural of datum. If I measure the weights of various suitcases at an airport (and why not?), the weight of each one is a datum. When I have finished, I have many data. I cannot have a 'piece of data', which is like having 'a slice of cakes'.

One datum, two or more data. Rant over.

There are two main sorts of data. Numerical data (which is sometimes called quantitative – not an easy word to say), such as age, height and weight can be written as numbers. Qualitative data – such as your favourite colour, brand of car or footballer – cannot.

Numerical data can be discrete or continuous. Discrete data refers to numbers that have specific values. The number of children in a family is either 2 or 3, not 2.8 or 2.004 – it has to be a whole number. Continuous data, on the other hand, can take any value. If I measure someone's height, it isn't limited to a whole number of centimetres. I could use more and more precise equipment and measure them to a very small fraction of a centimetre.

We can (and do) collect data about all sorts of things.

Businesses gather information about their customers to see what 'type' of person likes their product and to work out how to make it appeal to other 'types' of person. They use data about how their products are selling to decide how many of them to make, where to send them and how often.

Averages

Let's really go back to the good old school days and imagine we have the exam results for a class of 21 children, shown in percent.

75 63 27 48 92 56 65 63 88 39 46 73 75 63
67 50 49 72 78 64 79

The human brain tends to switch off when presented with a list of numbers, so we need to extract some meaningful but more concise information from the list. This is where averages can help. At school, you were probably taught about three kinds of average: mean, median and mode.

Mean

The first and most commonly used kind of average is called the mean (sometimes the arithmetic mean). In fact, it is so commonly used that most people mean 'mean' when they say 'average'. We work out the mean by adding together all the data and dividing by how many data there were. In this case, that means adding together all the scores and dividing by 21, as there were 21 children in the class.

This gives us: 1332÷21= 63.428571... which we'd probably round to the nearest percent. So the mean mark for the class would be 63%.

The mean can often be a value that does not match any of the data. Our mean was 63.428571, which no one actually scored. This throws up some interesting points: for instance, what is the mean number of arms each person has? Well, by far and away most people have two arms, but there are some people unfortunate enough to be missing one or both arms. These far outweigh the number of people with more than two arms. So, according to the mean at least, the average human has fewer than two arms.

Median

The median can help where the mean fails. The median is the middle datum in a set of data that has been ranged from lowest to highest. In our class of 21 examinees, we would need to order the results like this:

27 39 46 48 49 50 56 63 63 63 64 65 67 72
73 75 75 78 79 88 92

The middle mark out of 21 marks is the eleventh one

(it has 10 marks on either side of it). Some people find the middle mark by crossing off the data in pairs from each end:

~~27~~ ~~39~~ ~~46~~ ~~48~~ ~~49~~ ~~50~~ ~~56~~ ~~63~~ ~~63~~ ~~63~~ 64 ~~65~~ ~~67~~ ~~72~~
~~73~~ ~~75~~ ~~75~~ ~~78~~ ~~79~~ ~~88~~ ~~92~~

So the eleventh datum, 64%, is the median average. Unlike the mean, this is a mark that one of the children actually attained. The median number of arms for a human being is 2, because if we listed each person's quota of arms from lowest to highest, we would have relatively few 0s and 1s at the beginning and then the best part of 6 billion 2s. So we would be confident that the average human has 2 arms, according to the median.*

Mode

The mode is simply the most common datum. For our class of 21, putting the numbers in numerical order actually helps a lot with this, as it makes it easy to see that three children scored 63%, so the mode is 63%.

Sometimes there can be more than one mode. If the fellow who scored 73% had got an extra 2%, there would have been three pupils with 75% and our data would be bi-modal.

* If you have an even number of data, there won't be one middle value, but two – in which case you'd take the mean of those two numbers. And find yourself back in unrealistic territory . . .

There can also be no mode. If each child had a different score, no score would have occurred more than any other. Data with no mode is not the same as data whose mode is 0, which would have been the case if four or more children had scored 0 in the exam.

And if you're wondering about our average human's arms, the modal average is of course 2.

USING AND ABUSING AVERAGES

As you can see, because there are three ways of taking an average, different ones are selected according to what the statistician wants to put across. So when you see a newspaper headline claiming that 'the average Briton is missing an arm', find out what kind of average is being used before you take too much stock in it.

CHARTS

There are many types of chart. You'll see bar charts at stations, showing the performance and non-performance of various train lines, or pie charts on those political pamphlets that get shoved through your letterbox, showing that any problems are the other lot's fault, or scattergraphs all over the news, showing how far house prices have slumped this month.

Bar Charts and Pie Charts

The good thing about bar charts and pie charts is that they can be used to display non-numerical data as well as numerical data. Let's imagine you were asked to do a statistical survey at a big supermarket, in which you asked people which kind of wine they preferred. You could display your results in a bar chart, like this:

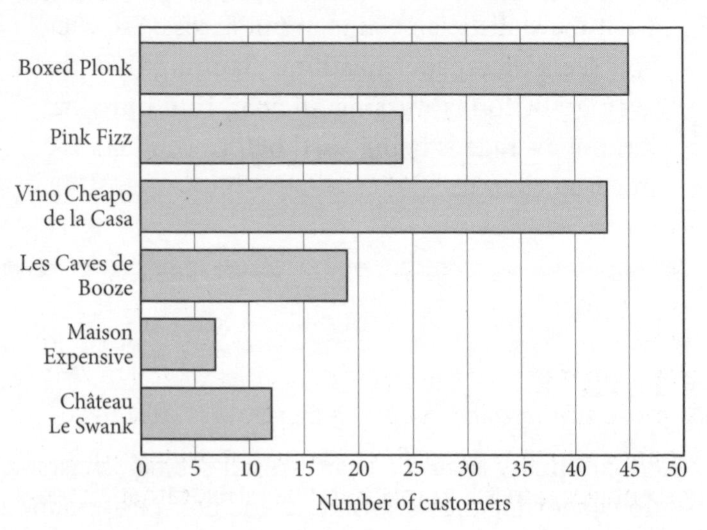

The relative lengths of the lines make it easy to see at a glance what's going on.

You could just as easily have the bars of your bar chart going vertically, with the numbers running up the left-hand axis, which is a fancy word for side.

And here's the same data presented in a pie chart, so named for obvious reasons:

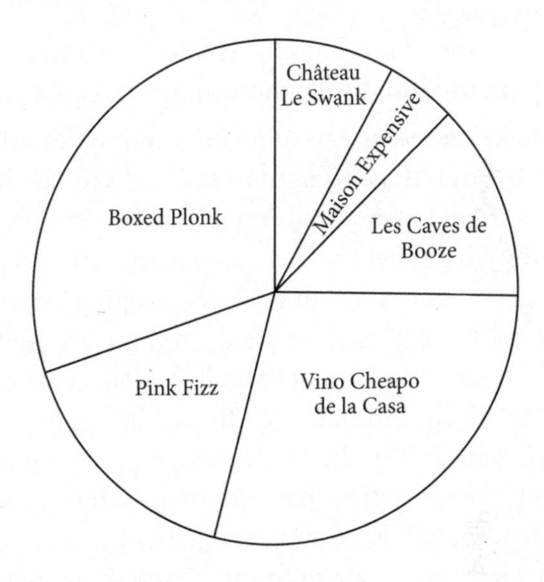

As with a bar graph, the pie chart shows the relative proportions of each kind of wine, but when you have two or more similar-sized slices, it can be less easy to see at a glance which is biggest. Often people would stick the percentages into the graph, but it can look rather a messy sort of pie.

Pie charts are trickier to draw than bar charts, because you have to figure out how many degrees each customer represents.* The survey has 150 customers in it and a circle has 360°, so each customer is represented by $360 \div 150 = 2.4°$. So the Boxed Plonk slice, which is 45 customers, would need to be $2.4 \times 45 = 108°$.

FIDDLY DATA

Sometimes, if you want to display lots of numerical data, it can make more sense to split your findings into slightly vaguer groups than to show each individual datum. Imagine if we'd also asked our 150 wine drinkers how much they had spent at the supermarket that day. We would get a wide range of answers, ranging potentially from 0 up to the person who did a month's shopping for a family of seven and bought a wide-screen television and the booze for his brother's wedding all in one go.

With 150 wildly different answers, we could still work out the mean, median and range of how much the customers spent, but it's unlikely that any two customers spent exactly the same amount. This means that there probably wouldn't be a mode – and if there were, it would be based on coincidence rather than meaning. And this is before we consider the tedium and impracticality of

---

* For more on degrees and angles, turn to p.140.

drawing a 150-bar chart or a 150-slice pie.

Far better to make a table and put the customers into groups (or classes, as they are called in maths), according to their spend, and then to use more manageable figures to create a simple bar chart:

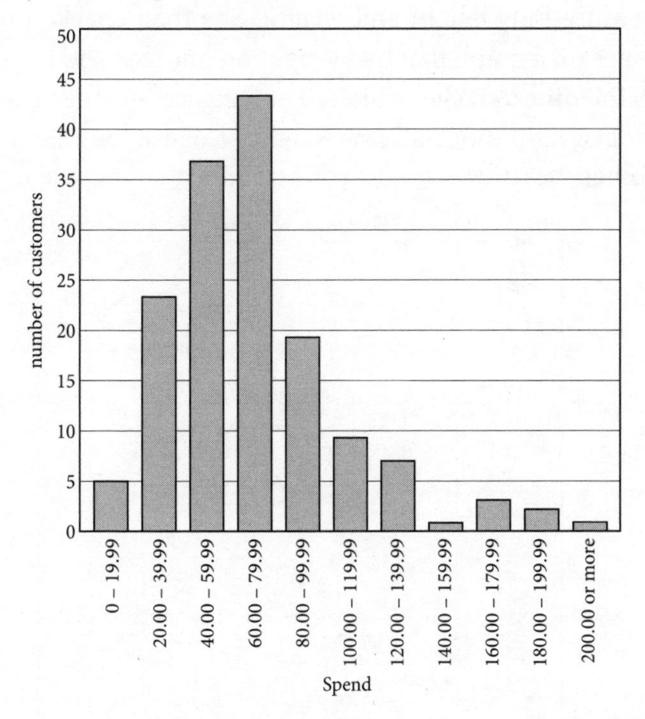

The 60–79.99 bar is the tallest, so this is the most common class: the modal class. Now we can stop loitering about outside supermarkets and think of other things.

Scattergraphs

Sometimes you have two pieces of data that are linked. When I registered with a new doctor, for example, the nurse asked me all sorts of embarrassing questions, and measured my height and weight. She then checked my values on a graph that had weight on one axis and height on the other, so she could tell at a glance whether I was underweight, normal, overweight or obese for a man of my height.*

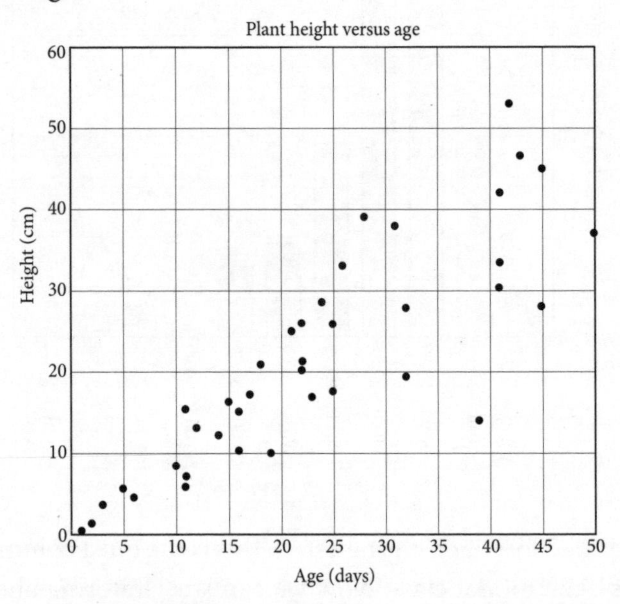

* She could presumably have worked this out just by looking at me, but that isn't terribly scientific, and it wouldn't illustrate my point.

A graph that shows the link between two sets of data (such as average height versus average weight) is called a scattergraph. Imagine I measured the height of some plants I'd been growing and also recorded how long each plant had been growing.

Each point represents a plant. The further each point is to the right, the older it is; the closer each point is to the top, the taller it is.

But why bother to draw this graph? I could have worked out the averages and ranges for the plants from my raw data – what more does the graph tell us?

Well, we can see that the points, broadly speaking, go from bottom left to top right. The graph therefore shows that, the older the plant, the taller it is. This is called a positive correlation and it's a way of spotting trends and relationships in data.

A negative correlation is one where the points go from top left to bottom right. For example, if I timed some children at an athletics club running 100 metres and compared it to their ages, I would expect to find their times decreased as they got older.

Scattergraphs can be useful as a means of predicting things. If the athletics club kept updating the data over a long period of time, they could use it as a way of predicting future performances from their youngsters. Scattergraphs are also a handy way of showing that a correlation exists. The government might publish a scattergraph showing cigarettes smoked per day versus age at death to try to put people off smoking, making it clear that there is a cause-and-effect situation in existence.

Of course, it is completely possible to have no correlation at all. If I compared women's heights to the number of times they've read *Pride and Prejudice*, I wouldn't expect any correlation.

Venn Diagrams

The last type of chart we're going to look at is a bit different from the others and perhaps doesn't belong in the statistics chapter at all: Venn diagrams.

WHO WAS THIS VENN AND WHAT WAS HE PLAYING AT?

John Venn was a British philosopher in the late nineteenth century who was very interested in logic. He came up with the diagrams (that are his namesake) in around 1880, as a way of showing how different sets of objects, numbers and even ideas are related.

Venn diagrams allow us to sort information and see relationships in a visual way, but using things called sets rather than data. Indeed, Venn diagrams are part of a field of mathematics called Set Theory, which has far-reaching implications but is nonetheless readily accessible to primary school children.

Let's look at an example. Before embarking upon a Venn diagram, we have to decide what our universal set is. Essentially: what is relevant to the survey? The universal set of this first example is going to be 'whole numbers from 1 to 10', but I might just as easily have chosen 'countries of the world', 'types of cheese' or 'things I want for Christmas'.

Traditionally, the universal set is drawn as a rectangular diagram, so I stick the members of my set – the numbers 1, 2, 3, 4, 5, 6, 7, 8, 9 and 10 – into a rectangle:

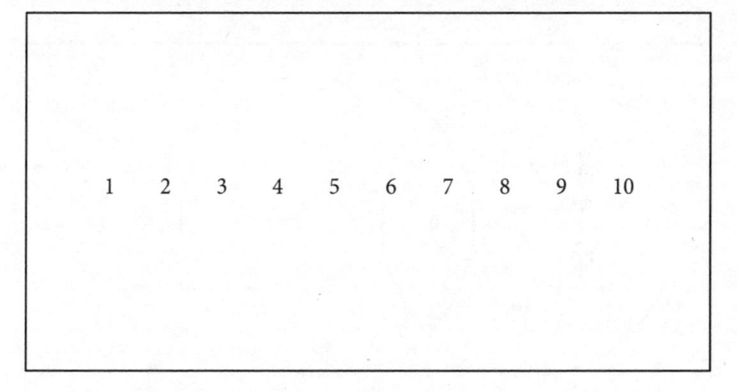

I'm now going to introduce some sets within my universal set. Set A will be the set of even numbers:

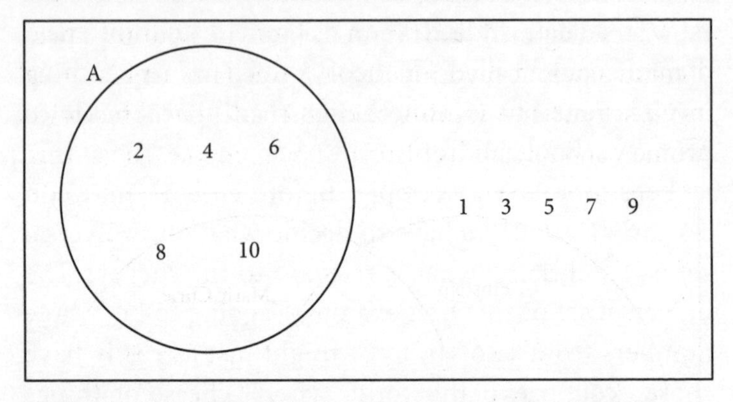

Next I add Set B, which I've decided will be the set of multiples of 3. But one number – 6 – is in both sets. This means we have to have Set A and Set B overlapping so that 6 can be in both at the same time:

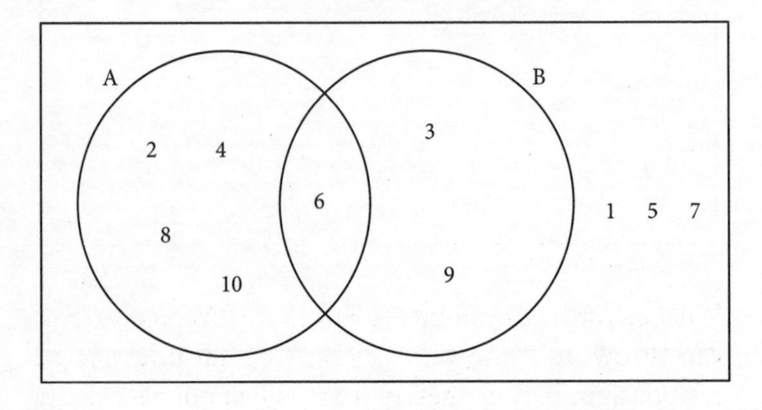

The overlapping part where the 6 lives is called the intersection of A and B.

Venn diagrams don't have to be done with numbers. I could say that my universal set was famous historical figures, that Set A was historical men, Set B was historical women and Set C was historical bald people. I'd get this:

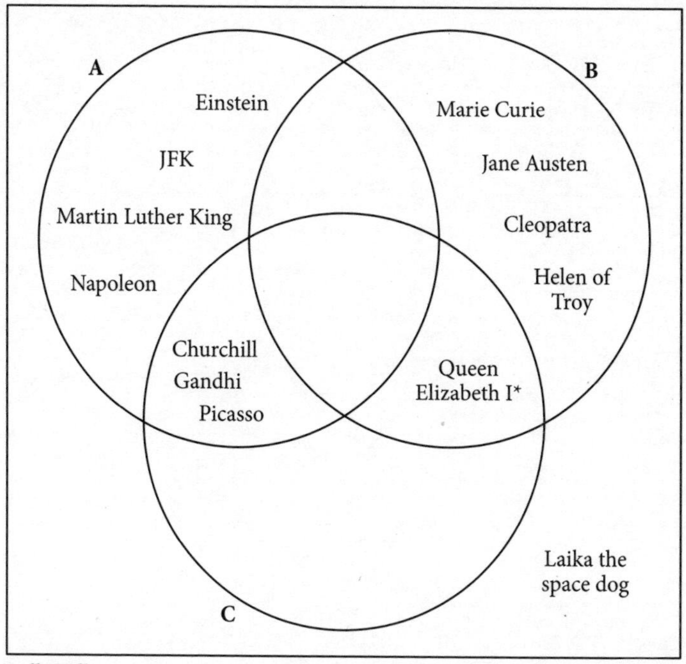

* allegedly

What on earth is the point of all this? Well, Venn diagrams can throw up some very interesting and unexpected relationships between sets of data – albeit not necessarily when we're dealing with bald queens and space dogs.

So that's statistics and common ways in which to display data. But how can we take measurements accurately in the first place? What systems of measurement do we have in place? The next chapter reveals all.

MEASUREMENTS

When the diverse and elusive rules of the mathematical universe (pure maths) come down to Earth and actually have to mean something (applied maths, science) it helps if there is some way to quantify things, a universal way of measuring various quantities. How long is a piece of string? How heavy is that baby? How hot is that latte?

SI UNITS

Anything that can be measured has a certain type of unit that goes with it. A good while back, a gang of (largely French) arch-boffins came up with standard units for all the things we can measure. These are called SI units (Système International d'Unités) and all mathematicians and scientists use them so that they can understand and use each other's results. This is basically an expanded form of the metric system.

The metric system is nice to work with because everything happens in powers of 10 and it's easy to convert from one metric denomination to another. If you are 190 centimetres tall, it's the work of a moment to divide by 100 and find that you are 1.9 metres tall. But if you use the imperial system and are 2 yards tall, but want to know how much that is in inches, you'd have to multiply by 3 to work out how many feet tall you are, and then multiply by 12 to get the number of inches, giving 2×3×12=72 inches, which is a little bit trickier.

The three most basic and useful SI units are:

Measurement	Unit	Symbol
Mass	Kilogram	kg
Time	Second	s
Length	Metre	m

Most other everyday measurements can be derived from these three units.

Prefixes

The SI system gives a basic unit for each type of measurement, which can be modified to indicate greater or smaller amounts by the use of certain prefixes such as mega-, kilo- or nano-. These prefixes help to keep the number part of the measurement manageably low: saying it is 17 kilometres to the next town is reasonable, while calling it 17,000,000 millimetres is less so.

Millimetres are very small units of length and are obtained by using the standard unit for length (metres) combined with the prefix milli- (meaning one-thousandth). But even a millimetre is large if for whatever reason you were trying to measure the width of an atom. Equally, a kilometre is a long way in shoes that rub, but it isn't at all handy for measuring the distances between stars.

Many of these prefixes are in everyday use, and even the larger ones are becoming more commonly used as computers get faster and offer more storage space:

Prefix	Name	Multiplier	Example
T	Tera	1 trillion	terabyte
G	Giga	1 billion	gigabyte
M	Mega	1 million	megatonne
k	Kilo	1 thousand	kilometres, kilogram
h	Hecto	1 hundred	hectares
c	Centi	1 hundredth	centimetres
m	Milli	1 thousandth	millilitres
μ	Micro	1 millionth	microscopic
n	Nano	1 billionth	nanotechnology

QUANTIFYING AND CONVERTING ALL SORTS OF THINGS

What follows is a list of the main things we might want to quantify and how their SI units work.

Mass

Mass is a measure of how much gravity affects an object. An object with a large mass needs a large force to lift it against the downward force of gravity, which means it's heavy.

Mass is not the same as weight, although we use the two terms interchangeably. Weight is (to a scientist) the force caused by gravity acting on the mass, rather than the mass itself. Hence an astronaut on the Moon – where gravity's pull isn't as strong as on Earth – weighs only one-sixth of what he weighs down here, although his mass remains the same.

Most common mass units happen in thousands:

1000 grams (g) = 1 kilogram (kg)
1000 kilograms (kg) = 1 tonne (t)

A gram is the mass of a grain of rice.
A kilogram is the mass of a one-litre bottle of water.
A tonne is the mass of a small car.

1 ounce = 28 grams
Quick conversion: Multiply the number of ounces by 3, then by 10
Calculator conversion: Multiply the number of ounces by 28

1 pound = 454 grams (or 0.454 kilograms)
Quick conversion: Multiply the number of pounds by

1000, and then divide by 2
Calculator conversion: Multiply the number of pounds by 454

1 stone = 6.35 kilograms
Quick conversion: Multiply the number of stones by 6
Calculator conversion: Multiply the number of stones by 6.35

1 ton = 1016 kilograms (or 1.016 tonnes)
Quick conversion: A ton and a tonne are similar enough that they can be used interchangeably

1 kilogram = 2.2 pounds
Quick conversion: Multiply the number of kilograms by 2
Calculator conversion: Multiply the number of kilograms by 2.2

1 tonne = 0.98 tons
Quick conversion: A tonne and a ton are similar enough that they can be used interchangeably

Length or Distance

The most commonly used units of length are, in order of increasing size: millimetres, centimetres, metres and kilometres.

10 millimetres (mm) = 1 centimetre (cm)
1000 millimetres (mm) = 100 centimetres (cm) = 1 metre (m)
1000 metres (m) = 1 kilometre (km)

A millimetre is about the thickness of a CD.
A centimetre is about the width of a pen lid.
A metre is the distance between the floor and most door handles.
A kilometre is the distance you'd cover in a fifteen-minute stroll.

1 inch = 2.54 centimetres (or 25.4 millimetres)
Quick conversion: Multiply the number of inches by 10, divide by 2, and then divide by 2 again
Calculator conversion: Multiply the number of inches by 2.54

1 foot = 0.305 metres (or 30.5 centimetres)
Quick conversion: Divide the number of feet by 3
Calculator conversion: Multiply the number of feet by 0.305

1 yard = 0.914 metres (or 91.4 centimetres)
Quick conversion: Divide the number of yards by 10, and then multiply by 9
Calculator conversion: Multiply the number of yards by 0.914

1 mile = 1.609 kilometres (or 1609 metres)
Quick conversion: Take the number of miles and add on half as many again, plus a tenth of your original number
Calculator conversion: Multiply the number of miles by 1.609

1 centimetre = 0.39 inches
Quick conversion: Divide the number of centimetres by 10, double your answer, and then double it again
Calculator conversion: Multiply the number of centimetres by 0.39

1 metre = 39.4 inches
Quick conversion: Double the number of metres, double your answer, and then multiply by 10
Calculator conversion: Multiply the number of metres by 39.4

1 kilometre = 0.62 miles
Quick conversion: Halve the number of kilometres, and then add on a tenth of the original amount
Calculator conversion: Multiply the number of kilometres by 0.62

AREA

Area describes the amount of flat space something takes up, such as the floor-space of an office. Area is normally described using units of length squared – for instance, a square metre is the area of a square with sides one metre long.

100 square millimetres (mm^2) =
1 square centimetre (cm^2)
100,000 square millimetres (mm^2) =
1 square metre (m^2)
1,000,000 square metres (m^2) =
1 square kilometre (km^2)

A square millimetre is about the area of a pinhead.
A square centimetre is about the area of the chip on a credit card.
A square metre is about the area of the bonnet of a small car.
A square kilometre is the area of two and a bit Vatican Cities stuck together.

A common mistake is for people to think that, because 100cm=1m, $100cm^2$ must be the same as a square metre. But picture that square metre: all of its sides are 100cm in length, so its area is $100cm \times 100cm = 10,000cm^2$. Likewise, a square centimetre is $10mm \times 10mm$, which gives $100mm^2$, and a square kilometre is 1000m×1000m, which gives $1,000,000m^2$.

1 square foot = 0.09 square metres
Quick conversion: Divide the number of square feet by ten
Calculator conversion: Multiply the number of square feet by 0.09

1 square yard = 0.84 square metres
Quick conversion: Divide the number of square yards by 10, then double your answer three times
Calculator conversion: Multiply the number of square yards by 0.84

1 acre = 4047 square metres
Quick conversion: Multiply the number of acres by 4, and then by 1000
Calculator conversion: Multiply the number of acres by 4047

1 acre = 0.4 hectares
Quick conversion: Divide the number of acres by 10, double your answer, then double it again
Calculator conversion: Multiply the number of acres by 0.4

1 square metre = 1.09 square yards
Quick conversion: Take the number of square metres and add on a tenth of that amount
Calculator conversion: Multiply the number of square metres by 1.09

1 hectare = 2.47 acres
Quick conversion: Divide the number of hectares by 4, and then multiply your answer by 10
Calculator conversion: Multiply the number of hectares by 2.47

VOLUME

Volume is used to describe how much space something takes up – but instead of dealing with flat surfaces (as area does), volume deals with 3D objects.* In everyday life, volume is normally used to describe an amount of liquid. We buy petrol by the litre, wine in 75-centilitre bottles and cola in 330-millilitre cans. We even talk about the size of a car's engine in terms of volume – litres or maybe even cubic centimetres.

If you imagine a cube whose edges are 1cm long each, that is a cubic centimetre, or 1 cm^3. If you now imagine that cube is made of water – that is a millilitre.

* See p.176 for formulae that'll help you work out the volumes of various solids.

10 millilitres (ml) = 1 centilitre (cl)
1000 millilitres (ml) = 100 centilitres (cl) = 1 litre (l)

The spoon that comes with children's medicine holds 5 millilitres.
A dessert spoon holds about 1 centilitre.
Two pints of beer make up a little over 1 litre.

If you are dealing with cm^3 rather than millilitres, it's worth remembering that a metre cubed is 100cm×100cm×100cm, which means that $1{,}000{,}000cm^3 = 1m^3$.

1 pint = 568 millilitres (or 0.568 of a litre)
Quick conversion: Multiply the number of pints by 6, and then by 100
Calculator conversion: Multiply the number of pints by 568

1 litre = 1.76 pints
Quick conversion: Double the number of litres
Calculator conversion: Multiply the number of litres by 1.76

Speed

We don't have a single unit for speed, in that we don't talk about a person or car moving 'at a speed of 9 zooms' or anything like that. Instead we talk about how far something goes in a given amount of time, which means that speed is an example of what is known as a derived unit.

While the metric system is becoming ever more popular in most aspects of life, speed is often still described in terms of miles per hour, rather than kilometres.

Meanwhile, scientists prefer to think about speed in metres per second, because both metres and seconds are basic SI units whose small size gives more accurate results.

The speed of sound is about 340 metres per second, or over 1200 kilometres per hour.
The speed of light is a shade under 300,000 kilometres per second, way faster than sound, which is why we see lightning before we hear the thunder, and why it looks as if rock stars are miming when you're sitting in the cheap seats at the back of the stadium.

1 mile per hour = 1.61 kilometres per hour
Quick conversion: Take the number of miles per hour and add half as many again, and then add a tenth of your original number
Calculator conversion: Multiply the number of miles per hour by 1.609

1 kilometre per hour = 0.62 miles per hour
Quick conversion: Halve the number of kilometres per hour, and then add a tenth of the original amount
Calculator conversion: Multiply the number of kilometres per hour by 0.62

TEMPERATURE

What we think of as the hotness or coldness of an object is down to the vibration of the molecules within it. The more energy – or heat – you put into an object, the more the molecules vibrate and the hotter that object becomes.

Most of the world happily uses the Celsius (sometimes called Centigrade) scale, a few countries (such as the US) use Fahrenheit, and everyone lets scientists use the Kelvin scale as long as they don't go on about it.

WHO WERE FAHRENHEIT, CELSIUS AND KELVIN?

Over the years, there have been various scales for measuring temperature. German physicist Daniel Fahrenheit came up with the temperature scale that bears his name in the early eighteenth century, using the temperature of his wife's armpit as a key value in his scale. A short while later, Anders Celsius established his own scale, with 0° as the boiling point of water and 100° as its freezing point – it took the intervention of a fellow Swede, Carl Linnaeus, to turn it 'the right way round'.

The trouble with these two scales is that they are not accurately linked to the molecular wobbling. While 2m is twice as long as 1m, 2° Celsius is not twice as hot as 1° Celsius.

Scotsman William Thomson, 1st Baron Kelvin, was the first person to think about temperature in terms of wobbling molecules, working out in the mid-nineteenth century that the vibrations stop completely at -273°C – this is Absolute Zero, the coldest it is possible to be. So 0 on the Kelvin scale is equal to -273°C, and converting between the two is nice and easy since Kelvin helpfully used the size of a Celsius degree as the template for the degrees on his own scale. If you want to work out how close we are to Absolute Zero in the middle of winter (answer: nowhere near), add 273 to the Celsius temperature: -10°C is equal to a scorching 263 Kelvin.

Water freezes (or ice melts) at 0°C.
Normal human body temperature is about 36–37°C.
Water boils at 100°C.

To convert from Fahrenheit to Celsius:

$$\text{Temperature}\,^{\circ}\text{C} = (\text{Temperature}\,^{\circ}\text{F} - 32) \times 5/9$$

To convert from Celsius to Fahrenheit:

$$\text{Temperature}\,^{\circ}\text{F} = (\text{Temperature}\,^{\circ}\text{C} \times 9/5) + 32$$

ALGEBRA

Algebra is the branch of mathematics in which letters stand for numbers. What's the point of that? Well, it helps with calculations and formulae in which one or more elements have an unknown value but a fixed relationship with the other values. For instance:

$$S=D\div T$$

This formula, in words, would be 'Speed equals Distance divided by Time'; S stands for speed, D for distance and T for time. There is a fixed relationship between those three elements that determines speed, and you simply need to stick in your own details in order to work it out. So if you have travelled 100km in 2 hours, your average speed would be:

$$S=D\div T$$
$$S=100\div 2$$
$$= 50\text{km/hour}$$

There are many formulae in maths and science that various people have worked out over the years. You plug the numbers into the formula, do the necessary maths, and out comes the quantity that you wanted to know.

E=MC²

Albert Einstein's $E=mc^2$ is the most famous bit of algebra in the world. We've all heard of it but what on earth does it mean?

It's all to do with the equivalence of mass and energy, which, you'll be pleased to hear, is far beyond the scope of this book. E stands for energy, m for mass and c for the speed of light in a vacuum. The speed of light is a huge number (300,000,000m/second), and the speed of light squared is enormous (300,000,000×300,000,000=90,000,000,000,000,000). So the energy you get from turning even a small bit of mass into energy is correspondingly large – see atomic bombs and nuclear reactors for a demonstration…

BASIC ALGEBRA

Algebra was not invented purely so that scientists could come up with famous equations. It is used for solving problems, be they in engineering, economics or maths exams. To solve the problems, the first step is to form an equation.

The difference between a formula and an equation is that the letters in a formula can take a whole range of values depending on the circumstances, whereas those in an equation can't – there tends to be a limited number of values (often only one) that make the equation hold true.

An equation is simply one or more bits of algebra linked by an = sign. Perhaps the most basic possible equation would be something along the lines of:

$$y=1$$

There's not a lot to grasp here: I have some unknown, y, which happens to equal 1. There's nothing to work out here so let's move on to something that involves some thought. Equations can – and do – get a lot more complicated than this.

The following problems are all quite easy, but it's less the answer that's important than our method of reaching it. Getting to grips with a few equation-solving rules will come in handy when we get to the ones we can't work out on our fingers.

Q1. I'm thinking of a number. If I add 5 to my number, I get 9. What is my number?

You don't need to hold a maths degree to see that the answer is 4, because 4+5=9. However, if I were to set this up as an equation, I'd start by giving the original number a letter, say Q.* I would then form an equation using the information I was given:

$$Q+5=9$$

Now, pretending that I don't already know the answer, I need to isolate Q on one side of the = sign, so as to work it out using a simple sum.

But how to get rid of the '+5'? I'm going to use the one and only (believe it or not!) rule for solving equations:

'As long as you do the same thing to both sides of an equation, the equation will still be true.'

I'm going to subtract 5 from both sides of my equation:

$$Q+5=9$$
$$Q+5-5=9-5$$

This may look a bit weird, but bear with me. Now that the left-hand section has '+5–5', the two 5s cancel each other out and I am left with some simple deductions:

$$Q+0=9-5$$
$$Q=4$$

---

* X is a popular choice, but I'm going to avoid that for the time being as it looks like a multiplication sign and I'm trying to make things clear.

It's an incredibly long-winded way of getting an answer to a problem we had already solved, but it does show a way to solve equations. I want to get my unknown quantity by itself, and can systematically eliminate anything that stands in my way.

> Q2. I'm thinking of a number. If I double my number and add 3, I get 15. What is my number?

You may well have figured out that the answer is 6, but I'm going to play dumb and form an equation. Doubling is the same as multiplying by 2, so if we randomly call the original number T, the equation would be:

$$2 \times T + 3 = 15$$

At this point, it's worth recalling BIDMAS (p.35), which dictates that the multiplication in this sum takes precedence over the addition. Because I can't do the multiplication yet, having no idea what T represents, I'll start off by getting rid of the '+3' in the same way as I did in the first example – by subtracting it from both sides:

$$2 \times T + 3 = 15$$
$$2 \times T + 3 - 3 = 15 - 3$$
$$2 \times T + 0 = 12$$
$$2 \times T = 12$$

Subtracting 3 from both sides neutralized that part of the equation and I can now deal with the '2×'. I effectively need to work out what half of 12 is in order to know what one

T is worth, so I divide both sides of my equation by 2:

$$2 \times T = 12$$

$$\frac{2 \times T}{2} = \frac{12}{2}$$

It's important that, when you do something to one side of the equation, you do it to the whole side. If I divided only 2 or only T by 2, I'd get the wrong answer.

The left-hand side is now 2 lots of T divided by 2, which leaves us with just the one T, while on the other side we can easily see that 12÷2=6, so:

$$T=6$$

Q3. I'm thinking of a number. If I divide my number by 3 and take away 4, I get 2. What is my number?

Calling the unknown number F for a change, this is our equation:

$$\frac{F}{3} - 4 = 2$$

First things first: we need to get rid of the '–4', this time by adding 4 to both sides:

$$\frac{F}{3} - 4 + 4 = 2 + 4$$

$$\frac{F}{3} = 6$$

Continuing our pattern of neutralizing unwanted elements

by doing their opposite, we now just need to multiply both sides by 3 in order to get rid of the '$\overline{3}$', and then we're done:

$$\frac{F}{3} \times 3 = 6 \times 3$$

$$F = 18$$

> Q4. I'm thinking of a number. Triple my number minus 4 is the same as double my number plus 3. What is my number?

A minor complication arises when I have an equation that has the unknown number on both sides of the = sign. Using R as our number, this is the equation:

$$3 \times R - 4 = 2 \times R + 3$$

Now is probably a good time to mention that, in algebra, a shorthand way of writing 3×R is simply 3R – a number next to an unknown means that they are multiplied. This turns our equation into the less unsightly:

$$3R - 4 = 2R + 3$$

I can't easily solve the equation while there's an R on both sides, so I need to eliminate one of the two sets of Rs. A good rule of thumb is to eliminate the smaller one, so I'll subtract 2R from both sides:

$$3R - 4 - 2R = 2R + 3 - 2R$$

$$\text{or } 3R - 2R - 4 = 2R - 2R + 3$$

This rather neatly leaves me with just one R on the left and none on the right:

$$R-4=3$$

We know what to do from here:

$$R-4+4=3+4$$
$$R=7$$

CHECKING YOUR ANSWERS

The good thing about solving equations – aside from the intrinsic satisfaction of doing it, says my inner maths geek – is that you can always check whether or not your answer is correct. Simply stick your answer back into the original equation in place and see if it works out:

$$3\times R-4=2\times R+3$$
$$3\times 7-4=2\times 7+3$$
$$21-4=14+3$$
$$17=17$$

SIMPLIFYING EXPRESSIONS

Each side of an equation is called an expression. The key to solving equations quickly and accurately is understanding how to simplify each expression – that is, make it as concise as possible before getting started. However, it can be very tempting to do too much when simplifying, especially when things get more complicated.

LIKE AND UNLIKE TERMS

Maths teachers often talk about like and unlike terms. Like terms are bits of algebra that have the same unknown (or unknowns) with the same powers, and unlike terms are ones that do not. For instance:

A and 3A are like terms	same unknowns with the same powers
A and B are unlike terms	different unknowns
A and A^2 are unlike terms	same unknowns, but different powers
AB and AB are like terms	same combination of unknowns with the same powers
A^2B and AB^2 are unlike terms	same combination of unknowns but with different powers

Like terms can be added and subtracted, whereas unlike terms cannot. It's rather like the difference between adding 10 apples and 5 apples and getting 15 apples, and adding 10 apples and 5 bananas and still just having 10 apples and 5 bananas. Unlike terms can't be added together or subtracted, but the like terms from the list above can, and are therefore easy to simplify:

$$A+3A=4A$$
$$AB+AB=2AB$$

Both like and unlike terms can be multiplied or divided. In the case of multiplying, we simply write the unknowns next to each other, which doesn't simplify things very much. With dividing, most of the time we write the terms as fractions:

$$A\times A=A^2$$
$$A\times B=AB$$
$$A\div A=1 \text{ (anything divided by itself = 1)}$$
$$A\div B=\frac{A}{B}$$

Simplifying Indices

So $A\times A=A^2$, in the same way as $2\times2=2^2$. When we're dealing with indices, it's useful to remember that even numbers with no index do actually have an index of 1, which we obviously don't bother writing as it makes no difference to the number. $A \times A = A^2$ is in fact $A^1 \times A^1 = A^2$, which shows quite nicely that, when we multiply using indices, we add the indices together:

$$Z^3 \times Z^4 = Z^{3+4} = Z^7$$
$$5^{23} \times 5^{12} = 5^{23+12} = 5^{35}$$

Note that this rule works only when the unknowns or numbers are the same.

If there are numbers attached to unknowns – these numbers are called coefficients, by the way, which always sounds impressive when used in conversation – just multiply them together in the usual way:

$$5E^2 \times 3E^4 = (5 \times 3) \times E^{2+4} = 15E^6$$

Dividing with indices works the opposite way around, as you'd expect. Instead of multiplying coefficients and adding indices, we divide coefficients and subtract indices. Here are a couple of examples to illustrate:

$$Q^7 \div Q^3 = Q^{7-3} = Q^4$$
$$15D^5 \div 3D^2 = (15 \div 3) \times D^{5-2} = 5D^3$$
$$H^3 \div H^8 = H^{3-8} = H^{-5}$$

Yep, you can have negative indices!

EXPANDING BRACKETS

Sometimes equations come with an expression that involves brackets. If the unknown you're trying to find is inconveniently trapped inside those brackets, you'll need to get rid of – or expand – the brackets.

To expand brackets, we have to multiply everything inside the brackets by whatever is immediately outside them – since, in algebra, things that are next to each other are multiplied together. Separating each bit with a + sign keeps things nice and simple.

$$5(C+3)$$
$$=(5\times C)+(5\times 3)$$
$$=5C+15$$

Initially at least, it looks as if I'm turning one pair of brackets into two, but those are just there to 'show my workings' and keep things clear.

We can check our expansion because it should work no matter what the value of C is, since we haven't been told what 5(C+3) is supposed to equal. So if I randomly decide that C=8, we should get the same amount for 5(C+3) as we do for 5C+15:

$$5(C+3)=5(8+3)$$
$$=5\times 11$$
$$=55$$
$$5C+15=(5\times 8)+15$$
$$=40+15$$
$$=55$$

When rearranging or expanding numbers within brackets, remember to keep the operation signs (+, – etc.) with the number or term on their right. This is especially important with negative numbers or terms. In this example, the minus sign in the first bracket 'belongs' to the Q (and once again, you need to separate the terms with a + sign):

$$-2(4-Q)$$
$$=(-2\times4)+(-2\times-Q)$$

As we work out the second bracket, we need to bear in mind that a negative times a negative makes a positive:

$$=-8+2Q$$

> Q5. I'm thinking of a number. My number minus 4, times 5, is equal to 3 times my number minus 2. What is my number?

I have to put the n-4 in brackets because the puzzle says that 'minus 4' happens before 'times 5':

$$5(n-4)=3n-2$$

Expanding the brackets gives:

$$(5\times n)+(5\times-4)=3n-2$$
$$5n-20=3n-2$$

And now we are in familiar equation territory. Subtract 3n from both sides:

$$5n-20-3n=3n-2-3n$$
$$2n-20=-2$$
$$2n-20+20=-2+20$$
$$2n=18$$
$$\frac{2n}{2}=\frac{18}{2}$$
$$n=9$$

Just like that!

FACTORIZING

Factorizing is the opposite of expanding brackets: it's making brackets. It's useful for making formulae easier to use and remember, and it also allows us to solve equations that have unknowns with powers.

First off, here's a simple example that just uses numbers:

$$10+15=25$$

I can rewrite this as:

$$(5\times2)+(5\times3)=25$$

Each bracket now contains a 5, meaning that 10 and 15 both have 5 as a factor. I can use this to factorize –

$$5\times(2+3)=25$$

– which I can rewrite as:

$$5(2+3)=25$$

So I've factorized 10+15 to give 5(2+3). You would never ordinarily do this when you're only dealing with numbers, but the skill can be very useful for solving more complex equations – especially quadratic ones, which we'll come to on p.128.

Here's an example with an unknown. If my expression is 8A+10, both terms share a factor of 2. So I'm going to have 2 on the outside of my brackets:

$$8A+10=2(\ldots+\ldots)\ ?$$

Now, the way to fill in the bracket is to imagine you are expanding the bracket in reverse. I know that 2 times the first blank has to give me 8A, while 2 times the second bracket has to produce 10:

$$8A+10=2(4A+5)$$

You can also factorize when the expression has unknowns with powers in it. With $6B^2+9B$, both terms have a factor of 3, but they also both have B in them:

$$6B^2+9B=3B(\ldots+\ldots)\ ?$$

The first blank times 3B must give $6B^2$; $3B\times2B=6B^2$. The second blank times 3B must give 9B; 3B×3=9B.*

$$6B^2+9B=3B(2B+3)$$

We can also have situations where one of the terms in the problem is the factor, such as in $12y-24y^2$, in which the common factor is 12y. In these instances, in order to make a bracket, we need to remember that 12y multiplied by 1 is the same as 12y:

$$12y-24y^2=12y(1-2y)$$

A good rule of thumb is that we can always take out a factor (i.e. put it in front of the bracket) that is the problem's lowest-power unknown. If we have t^3+5t^2, for instance, we can use t^2 as the factor:

$$t^3+5t^2=t^2(t+5)$$

You can also factorize expressions with more than one unknown:

$$5xy+10x^2y^3=5xy(1+2xy^2)$$

And finally, you'll be relieved to hear that some problems don't have a common factor other than 1, in which case there's no point at all in factorizing:

$$5x+7y=1(5x+7y)$$

* N.B. It isn't 3B×3B because the B×B part would create B^2.

QUADRATIC EQUATIONS

All the equations we've solved so far have had unknowns that have been multiplied, divided, added to or subtracted from by a number. These are known as linear equations. You may well recall, perhaps with some degree of horror, another type of equation, known as quadratic. Quadratic equations are ones in which the unknown is squared, and can be much harder to solve.

Consider this:

$$y^2=4$$

This is saying that some number, y, times itself gives 4. I could write this as:

$$y \times y=4$$

Which number times itself equals 4? Well, obviously it's 2. But there's another answer as well... As we saw on p.39, multiplying a negative by a negative creates a positive, so it's also true that $-2\times-2=4$. So our equation – and indeed any equation that has a squared number in it, i.e. a quadratic equation – has two answers.

To solve my equation properly, doing the same thing to both sides, I have to square root both sides:

$$y^2=4$$
$$y=\sqrt{4}$$
$$y=\pm 2$$

The ± stands for 'positive or negative', and is a shorthand way of writing '2 or -2'.

That equation was very simple, but as soon as we add another term that has y in it, rather than y^2, we can't solve it in the same way as a linear equation. Take:

$$y^2+y=12$$

The trouble with solving this is that I can't combine the y^2 and the y into one thing – I can't do a series of things to both sides of the equation that will leave me with just y=something. If I tried subtracting y from both sides to get rid of the y, I'll get:

$$y^2+y-y=12-y$$
$$y^2=12-y$$

This doesn't help at all. Now I have y things on both sides of the equation, and still no clue what y actually is.

This is where factorizing helps. Rearrange the equation:

$$y^2+y=12$$
$$y(y+1)=12$$

By turning what was effectively $(y\times y)+(y\times 1)$ into $y(y+1)$, I discover that $y\times(y+1)=12$. If I make a guess that $y=3$, and that $y+1$ therefore is 4, the equation works.

To double-check the answer, let's insert this value into the original equation:

$$y^2+y$$
$$=3^2+3$$
$$=9+3$$
$$=12$$

So that's one of the solutions sorted, but what about the other? Well, it's really tricky to spot, but if we try y=-4, we'll get:

$$y^2+y$$
$$=(-4)^2-4$$
$$=16-4$$
$$=12$$

Hurray! But solving quadratic equations through trial and error is not very satisfactory. We need a technique that will lead us to both answers without too much brain-wracking over strange negative solutions. Is there such a thing? Of course there is, but it relies, once again, on factorizing and on modifying the original equation slightly.

Factorizing Quadratic Equations

To help you visualize how a quadratic expression can be factorized, take a look at the rectangle below. The longer side is y+5 metres long and the shorter side is y+3 metres long. I've split it into four smaller rectangles to show this, and to cunningly isolate the ys.

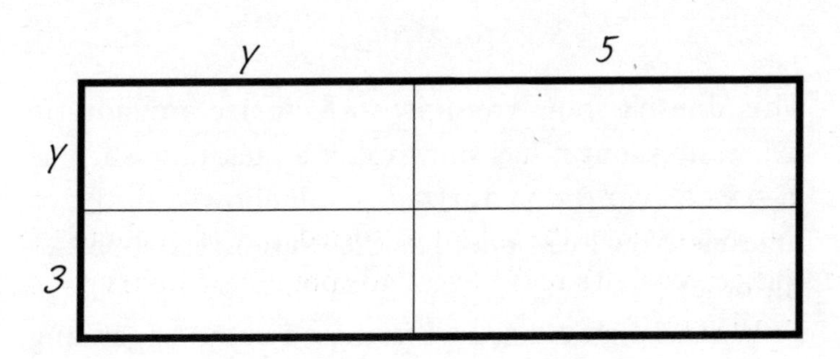

I can then work out the area of each of those rectangles:

	y	5
y	$y \times y = y^2$	$5 \times y = 5y$
3	$y \times 3 = 3y$	$5 \times 3 = 15$

So the area of the whole thing must be these parts added together:

$$y^2+5y+3y+15$$
$$=y^2+8y+15$$

However, we could also have got the area of the rectangle by multiplying the sides together. The sides are (y+3) and (y+5), so the area of the large rectangle, which we have just defined as the quadratic expression $y^2+8y+15$, must also be:

$$(y+3)(y+5)$$

This doesn't show you how to factorize a quadratic expression, but it has shown that a quadratic such as $y^2+8y+15$ can be factorized – somehow – into two brackets. Now we just need to know how to turn one into the other...

When I first started wittering on about expanding brackets, I said that, when you expand a bracket, you have to multiply everything inside the bracket by the thing outside the bracket. Here we have a bracket outside the bracket and it all looks rather confusing, so it helps to rearrange our two brackets into a less daunting format:

$$\begin{aligned}(\mathbf{y+3})(y+5) &= \mathbf{y}(y+5)+\mathbf{3}(y+5)\\ &= (y\times y)+(y\times 5)+(3\times y)+(3\times 5)\\ &= y^2+5y+3y+15\\ &= y^2+8y+15\end{aligned}$$

Which is back to where we started.

If we look closely, though, a clue as to how to do this backwards – that is, how to factorize the quadratic – is staring us right in the face. The brackets we worked out simply by looking at the large rectangle are both y plus some number. Looking at the quadratic, those two numbers have combined through addition to make 8 and through multiplication to make 15. This is the key to factorizing a quadratic: there is only one pair of numbers that make both an 8 and a 15: 3 and 5. Now we know which numbers will need to be in the brackets along with the ys.

Let's look at a few more examples. First off:

$$h^2+7h+10=(h+\ldots)(h+\ldots)\ ?$$

This is my starting point. I now need to think of two numbers with a sum of 7 and a product of 10 to fill in the blanks. I could draw up a list of possibilities to help:

First number	Second number	Sum	Product
1	6	7	6
2	5	7	10
3	4	7	12
4	3	7	12
5	2	7	10
6	1	7	6

There are two entries on my table that work: 2 and 5 or 5 and 2, which shows that it doesn't matter which way round I put the numbers in the brackets – the answer will be the same.

$$h^2+7h+10=(h+2)(h+5)$$

Another very quick example:

$$b^2+12b+35=(b+5)(b+7)$$

In this case, the numbers we needed were 5 and 7, because $5+7=12$ and $5\times7=35$.

Factorizing with Negatives

Things can get (even) trickier when you get combinations of positive and negative terms in your quadratic equation. We just worked out that $b^2+12b+35=(b+5)(b+7)$, but if, for the sake of making everything needlessly confusing, I change the +5 to a -5, I get:

$$b^2+2b-35=(b-5)(b+7)$$

This is because $-5+7=2$ and $-5\times7=-35$.

If I make the 7 negative, I get:

$$b^2-2b-35=(b+5)(b-7)$$

This is because $5-7=-2$ and $5\times-7=-35$.

If I make both the 5 and the 7 negative, I get:

$$b^2-12b+35=(b-5)(b-7)$$

This is because $-5-7=-12$ and $-5\times-7=35$.

Actually Solving Quadratic Equations

We still haven't actually solved any quadratic equations; we've just factorized a whole load of quadratic expressions.

The way to solve quadratic equations is to make one side of the equation 0. So let's say I'm asked to solve:

$$d^2+4d+2=7$$

My first move would be to make the right-hand side of the equation 0, by subtracting 7 from both sides:

$$d^2+4d+2-7=7-7$$
$$d^2+4d-5=0$$

Now I factorize. I'm looking for two numbers with a sum of 4 but a product of -5. A bit of head-scratching and I come up with -1 and 5:

$$(d-1)(d+5)=0$$

At this stage, I can exploit the fact that the only way to multiply two things together and arrive at 0 is if one of those two things is itself equal to 0. Any number in the world, when multiplied by 0, produces 0. In terms of my quadratic equation, this means:

$$\text{either } d-1=0, \text{ so } d=1$$
$$\text{or } d+5=0, \text{ so } d=-5$$

I can check this conclusion by substituting both 1 and -5

into my original problem in turn:

$$d^2+4d+2=7$$

$$\text{If } d=1$$
$$1^2+(4\times1)+2=7$$
$$1+4+2=7$$
$$7=7$$

$$\text{If } d=-5$$
$$(-5)^2+(4\times-5)+2=7^*$$
$$25-20+2=7$$
$$7=7$$

SIMULTANEOUS EQUATIONS

More equations? Don't worry, we're nearly there.

As well we all know, life is complicated. Often when mathematicians, engineers or scientists are trying to find the best way to sort out a real-life problem, they find that they actually end up with two or more problems – or indeed equations – that need to be satisfied at the same time. It's this 'at the same time' part that gives them their name: simultaneous equations.

---

* I put the -5 in brackets when I was squaring it, to remind myself to do $-5\times-5=25$ rather than $-(5\times5)=-25$.

Here's a little puzzle for you:

Q6. The combined age of my son and me is 37 years. I am 27 years older than him. How old are we?

You may well be able to work out the answer in your head by trial and error, but let's do it properly with a pair of equations. If I say that my age is given by the letter c and my son's age is given by the letter r, we get:

$$c+r=37 \text{ (our combined ages make 37)}$$
$$c-r=27 \text{ (I am 27 years older than him)}$$

These are simultaneous equations. The first one has an infinite number of solutions for c and r because, no matter what number you think of for c, there will always be a number for r (perhaps negative) that will bring the total up to 37. The same is true for the second equation when it is taken by itself. However, if both the equations have to be true simultaneously, there is only one possible answer for each of c and r.

There are several ways to solve simultaneous equations, but perhaps the most straightforward involves adding or subtracting the equations so that one of the unknowns gets eliminated. If I add my two equations together I get:

$$\begin{array}{rcccl} & c & + r & = & 37 \\ + & c & - r & = & 27 \\ \hline & 2c & & = & 64 \end{array}$$

The r terms have been eliminated because $r+-r=0$. I'm left with:

$$2c=64$$
$$c=32$$

So we've figured out that my age is 32.

To find my son's age, we just have to use the fact that $c=32$ in one of the original equations:

$$c+r=37$$
$$32+r=37$$
$$32+r-32=37-32$$
$$r=5$$

Now we know that I am 32 and my son is 5. These are the only values for c and r that work for both equations at the same time.

In that problem, we had only one lot of each unknown in each equation. Let's look at what we might do if this weren't the case:

$$5a+2b=16$$
$$4a-b=5$$

There's no way I can add or subtract these equations that totally eliminates either a or b.

But remember our golden rule of equations: as long as I do the same thing to both sides of the equation, the equation holds true. So if I simply multiply the second equation by 2, I get:

$$5a+2b=16$$
$$8a-2b=10$$

If I add them together now, the +2b and the -2b will eliminate, leaving:

$$13a=26$$
$$\frac{13a}{13}=\frac{26}{13}$$
$$a=2$$

To check my answer, I can substitute it into the original second equation, 4a–b=5:

$$4\times2-b=5$$
$$8-b=5$$
$$b=3$$

GEOMETRY

The word geometry comes from Ancient Greek and means 'Earth measuring'. Geometry investigates the relationships between lines, angle and shapes, and can be done in two or three – or more – dimensions.

What many people like about geometry is that it's very visual, and ancient geometers such as Euclid loved to prove their geometrical theories with diagrams rather than with calculations. It is this elegance that attracts many people who loathe other, more numerical areas of mathematics. That and the fact that you get to do lots of drawing and use exciting sharp-pointed compasses and pleasingly semicircular protractors.

ANGLES

An angle is formed where two lines meet or cross. We tend to use degrees to measure the size of an angle, which can be anywhere between 0° and 360°. A complete revolution is represented by 360°, apparently because the ancients worked out that it took the Earth roughly this long to complete an orbit of the Sun, so in each twenty-four-hour period the Earth moved 1° round the Sun. Pretty good work, I say.

There are various words that are used to describe different sizes of angle:

* **Acute**: angle x is less than 90°

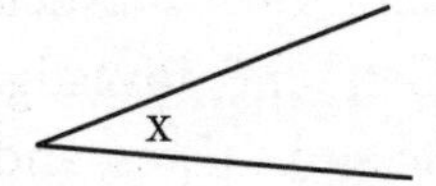

The word acute comes from the Latin for 'sharp'.

* **Right**: angle x is 90°

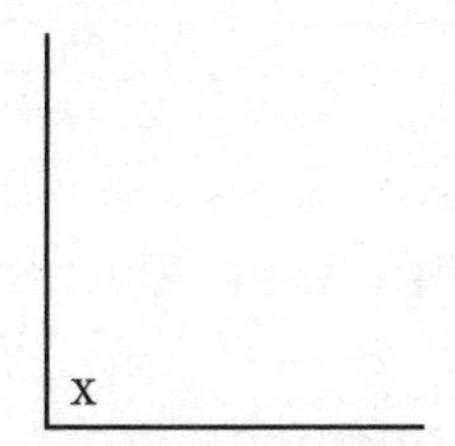

If you're building a house, you want a right angle between the floor and the wall. Lines at right angles are also called perpendicular.

* **Obtuse**: angle x is between 90° and 180°

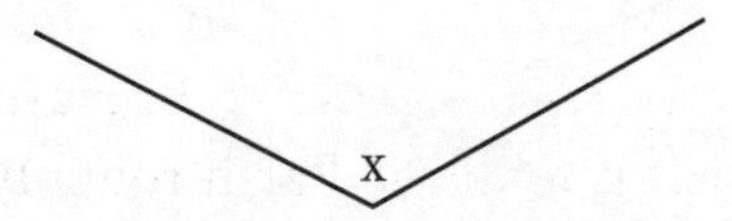

I shall remember to my dying day the pupil of mine who mistakenly wrote 'obese' rather than 'obtuse'. But you can see where he was coming from.

* **Straight line**: angle x is180°

x

If two lines meet head-on, they make 180°.

* **Reflex**: angle x is more than 180°

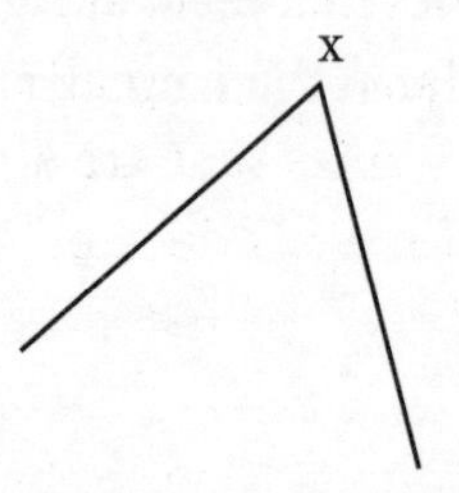

Vertically Opposite Angles

When two lines meet or cross they make what mathematicians call a vertex: a very sci-fi-sounding word for what everyday folk simply call a corner or a crossing. It is easy to see that angles opposite each other across a vertex are equal:

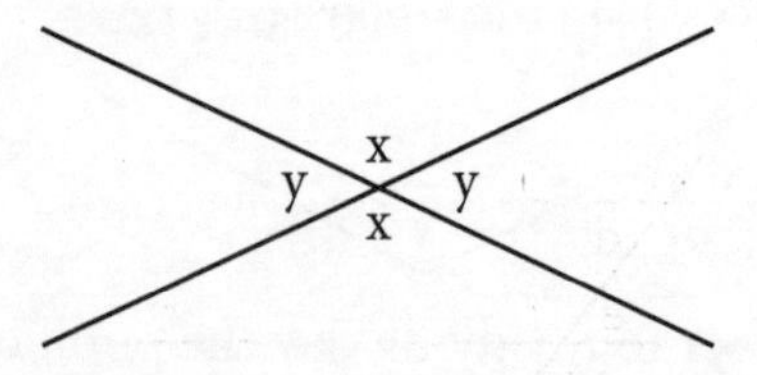

Confusingly, even for maths – and that's saying something – the adverb from vertex is vertically, nothing to do with the other vertically that means straight up or perpendicular to horizontal. So vertically opposite means 'on opposite sides of a vertex'.

Parallel Line Facts

Perpendicular lines – ones that meet at right angles – are pretty straightforward and crop up all over the place. Parallel lines, however – ones that are always the same distance apart and will never cross, even if they were extended infinitely – throw up some interesting angle scenarios.

Parallel lines are denoted with little chevrons:

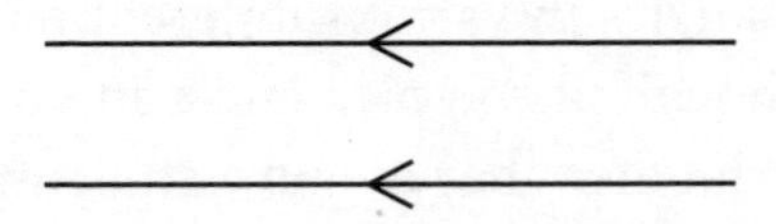

Now let's see what happens if we insert a third line that crosses (or transects) the parallel lines. We now have two vertices (plural of vertex) and eight angles, labelled a to h:

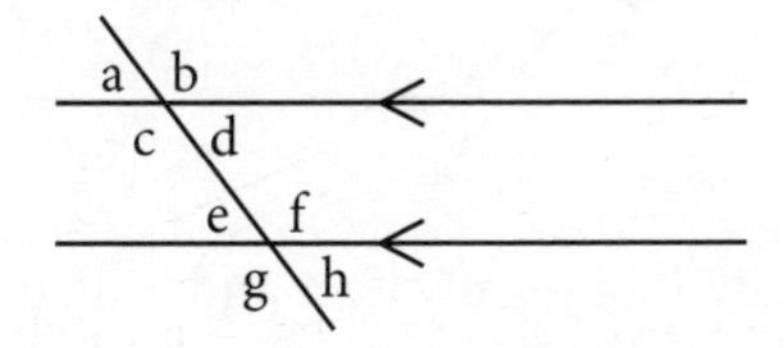

Coming back to our vertically opposite angles, we can see that a=d, b=c, e=h and f=g. Looking harder, you can see that the two vertices are identical, and the lines cross at the same angle. The a angles are in the same positions as the e angles and likewise with the bs and fs. They are called corresponding angles and are equal, so our diagram can be simplified even further:

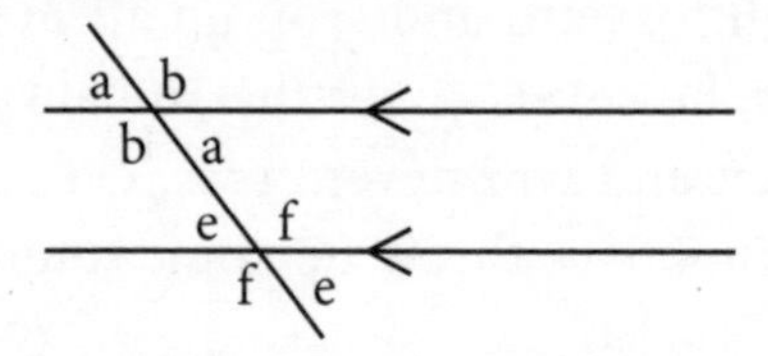

From the eight angles we started with, we now have only two different angles in various different places. It gets even better, though: if you pick any a and b angles that are next to each other, they are on a straight line and so must add up to 180°. For instance, if a (which is acute) is 50°, b would have to be 130° because 50°+130°=180°.

Consider angles p and q:

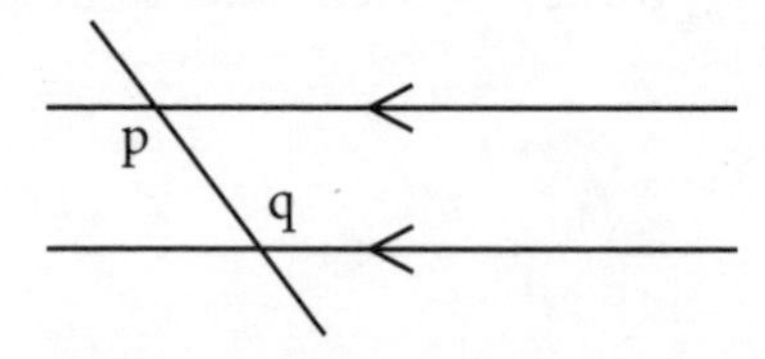

You'll notice, by looking at the a and b diagram we had above, that p and q are equal. Angles in this configuration are called alternate angles, and they are always equal.

Angles x and y are what we call co-interior angles, which always add up to 180°.

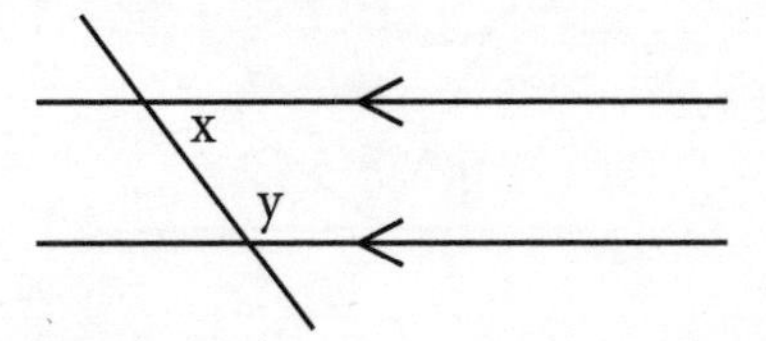

TRIANGLES

As you do more geometry, it is inevitable that the lines will eventually join up to form a closed shape – what mathematicians call a polygon, which is Ancient Greek for 'many angles'.

The simplest polygon, the one with the smallest possible number of lines, is the triangle. Triangles are very important because there are certain indisputable mathematical facts about them, which means that the easiest way to work out a non-triangle's geometrical information is to divide it up into triangles.

The first fascinating triangle fact is that all triangles, no matter what shape or size, have three interior angles that add up to 180°.

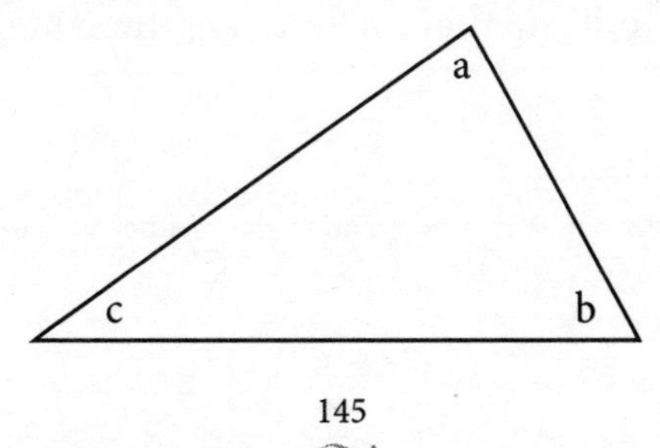

If I were to cut the corners of this – or any other – triangle off, I could fit them together along a straight line, and straight lines, as we've already seen, always add up to 180°:

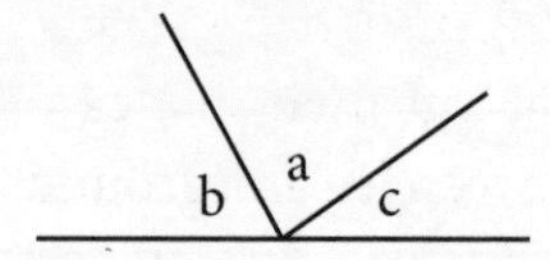

Types of Triangle

There are three types of triangle.

An equilateral triangle is one whose sides are all the same length and whose angles are all the same size. Since its identical angles add up to 180°, each one measures 180°÷3=60°.* An equilateral triangle looks like this:

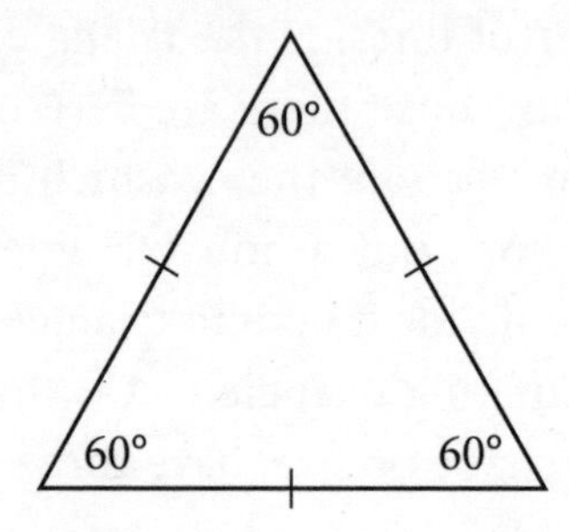

The dashes indicate that the lines are the same length.

* When shapes have equal sides and angles, mathematicians call them regular – more of which on p.153.

An isosceles triangle has two sides of the same length, and therefore two angles of the same size. Isosceles means 'equal legs', so I always think of isosceles triangles as being like a person standing upright on a flat surface. Their legs are the same length but their feet can be any distance apart. The angle between leg and ground will be identical on both sides:

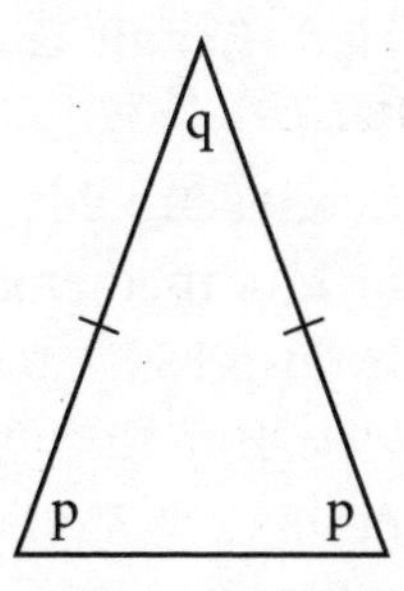

Unlike with the equilateral triangle, we can't say what the values of p and q are because they could be anything, so long as the angles add up to 180°. What we can say is that $2p+q=180°$.

The third species of triangle has no equal sides or angles, and is called a scalene triangle.

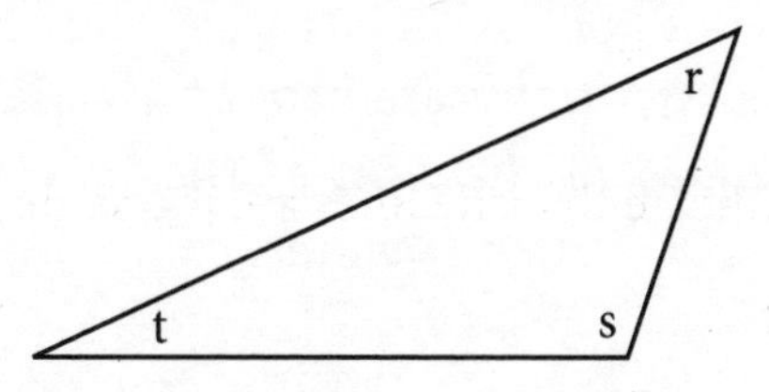

All we can say here is that $r+s+t=180°$.

Right-Angled Triangles

Right-angled triangles are an important sub-section of triangles. As their name suggests, they are triangles that contain a right angle: 90°.

It's a common error to think that these triangles constitute a fourth variety of triangle, but if you think about it, a right-angled triangle is necessarily either an isosceles or a scalene.

Right-angled triangles are very important in their own right and a branch of mathematics – trigonometry, which we'll come to on p.180 – has its foundations in right-angled triangles. But first, there's another vital mathematical subject that relies on them…

Pythagoras' Theorem

Pythagoras was another Ancient Greek scholar whose name has lived on in connection with a theorem to do with right-angled triangles. Pythagoras' Theorem states that:

'The square on the hypotenuse of a right-angled triangle is equal to the sum of the squares on the other two sides.'

What does this mean? Well, the hypotenuse is the longest side of a right-angled triangle and is always located opposite the right angle. The square on the hypotenuse is just that: a square drawn with the hypotenuse as one of its sides, square A in the diagram below:

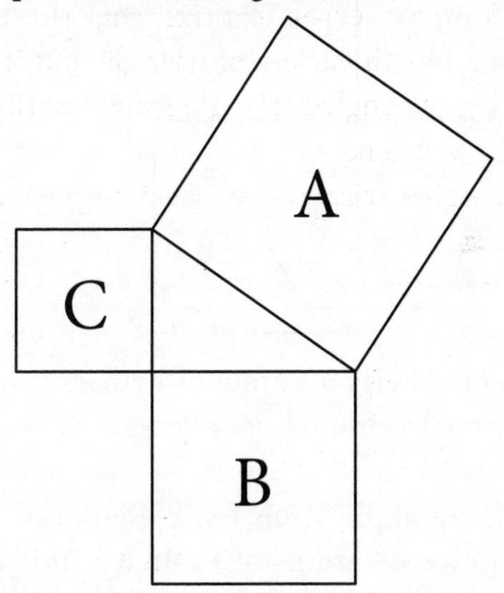

Squares B and C are the squares on the other two sides. So Pythagoras' Theorem says that the area of square A is equal to the areas of squares B and C added together, and this is true for whatever right-angled triangle we start with. You can see this on the following diagram:

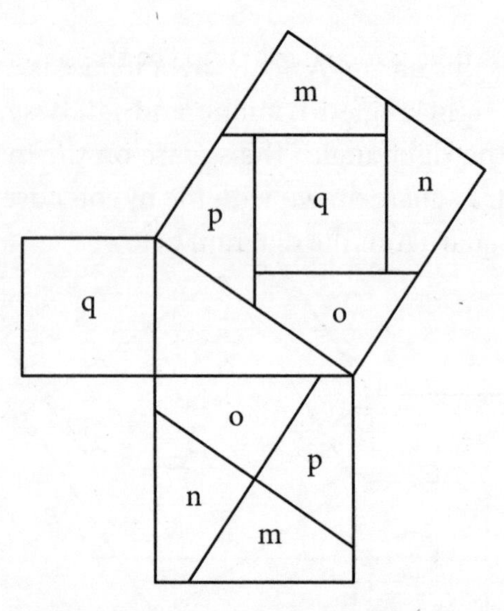

This particularly elegant proof of Pythagoras' Theorem was figured out by Briton Henry Perigal in the nineteenth century.

There is an application for Pythagoras' Theorem beyond strange diagrams of triangles and squares. It allows us to work out the length of the third side of a right-angled triangle if we know the other two.

Here's a triangle:

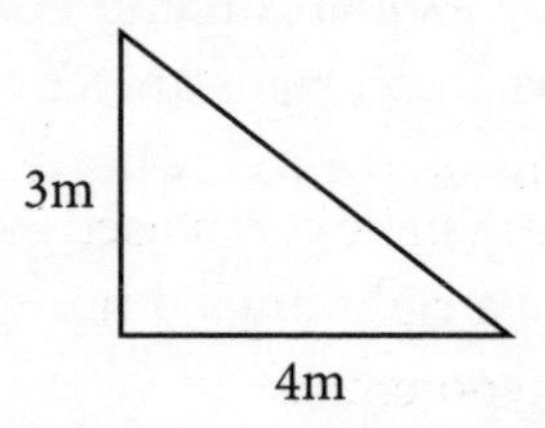

We know the lengths of two of the sides, but not the length

of the third one, which happens to be the hypotenuse.

The area of each square would be the length of each side squared. If I say that the length of the hypotenuse is h:

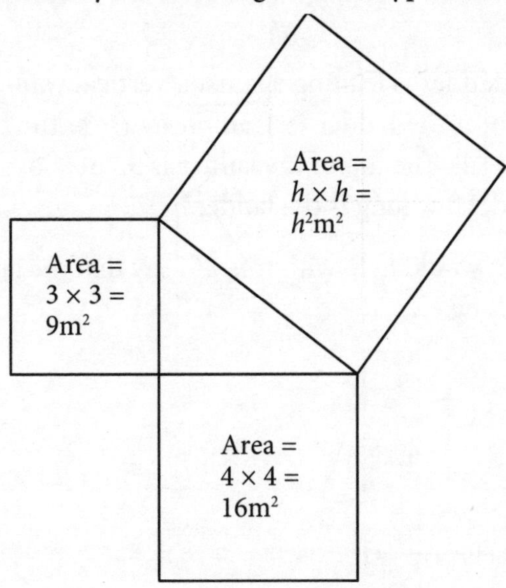

I know from Pythagoras' Theorem that $h^2=9+16$, so I just need to solve the equation:

$$h^2=25$$
$$h=\sqrt{25}$$
$$h=5$$

Pythagoras' Theorem can be condensed into a handy little equation. If we have a right-angled triangle whose sides are a, b and h (the hypotenuse):

$$h^2=a^2+b^2$$

This equation can be rearranged and solved, as long as you know the lengths of any two sides.

Here's a classic example of a Pythagoras-related question:

> Q1. A ladder is leaning against a vertical wall. The bottom of the ladder is 1.5m away from the base of the wall. The top of the ladder is 3.7m above the ground. How long is the ladder?

A diagram would help with this. If I say that the ladder is L metres long:

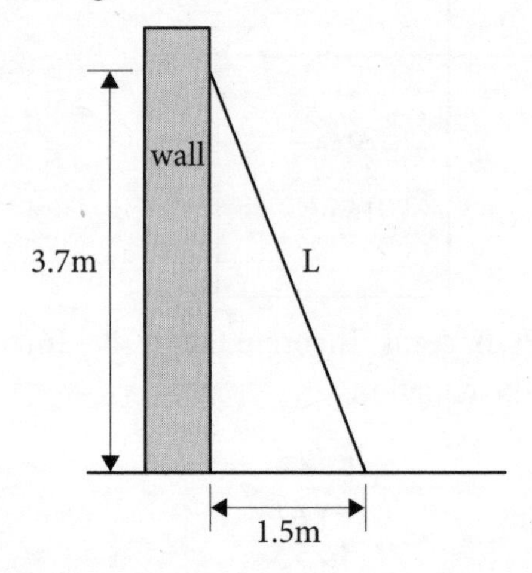

The ladder, wall and floor make a right-angled triangle, so I can use Pythagoras' Theorem. L is the hypotenuse as it's opposite the right angle.

$$L^2=1.5^2+3.7^2$$
$$L=\sqrt{(1.5^2+3.7^2)}$$
$$L=\sqrt{(2.25+13.69)}$$
$$L=\sqrt{15.94}$$
$$L=3.992492956$$

So it would seem sensible to give our final answer as L=4m.

POLYGONS OF ALL SHAPES AND SIZES

The next family of polygons are the ones with four sides, quadrilaterals.

A regular quadrilateral – and remember: regular means having equal sides and angles – has to be a square, because all its sides are the same length and it has four right angles. The angles of any shape that consists of four right angles necessarily add up to 360° (4×90°). Any quadrilateral can also be cut across diagonally to make two 180° triangles:

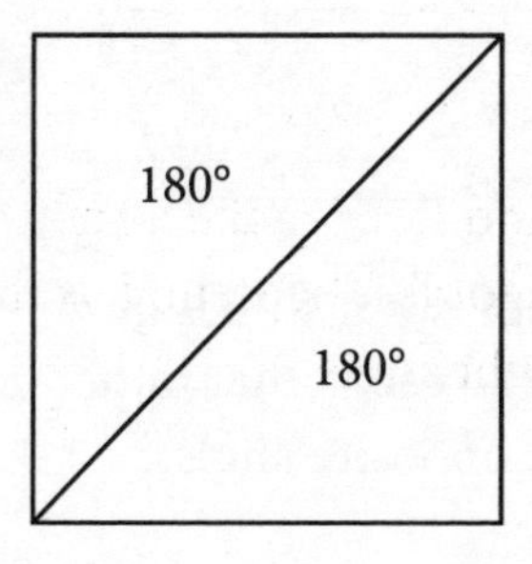

I don't need to explain what squares and rectangles are, so let's move on to some slightly more bizarre shapes. A slanted rectangle is a parallelogram, which also has two pairs of equal sides, but also two pairs of equal angles rather than four right angles:

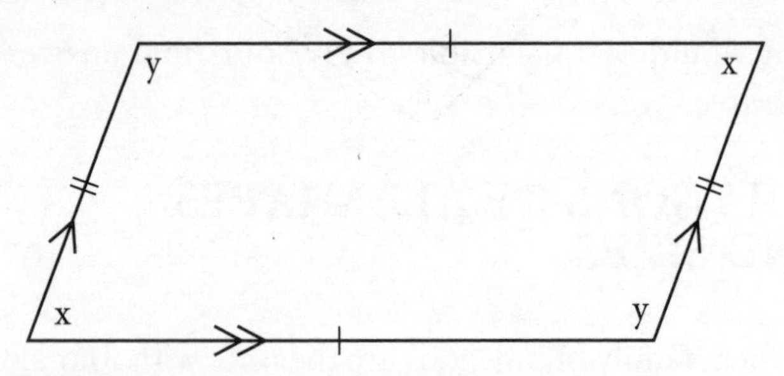

As we saw earlier in this chapter, x and y are co-interior angles, so $x+y=180°$, or indeed $2x+2y=360°$.

A rhombus is rather like a parallelogram, except that all of its sides are equal:

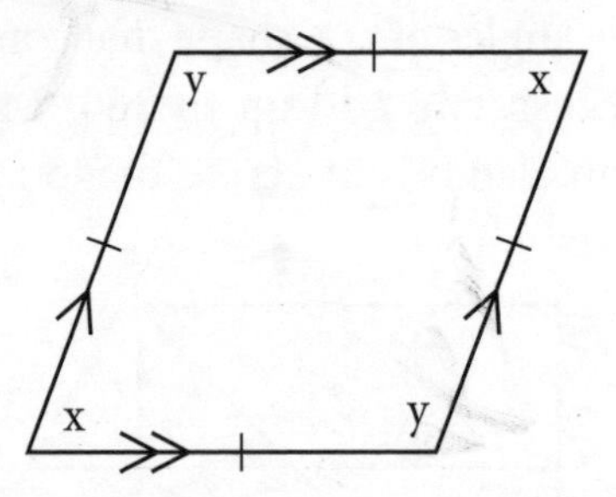

And then there are kites, which have two pairs of equal sides that are next to each other rather than opposite, and one pair of equal, opposite angles:

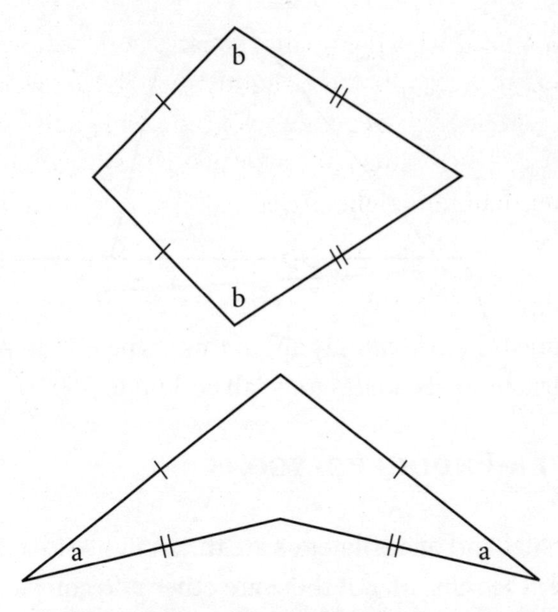

The last quadrilateral with a fancy name is a trapezium, which has just one pair of parallel sides:

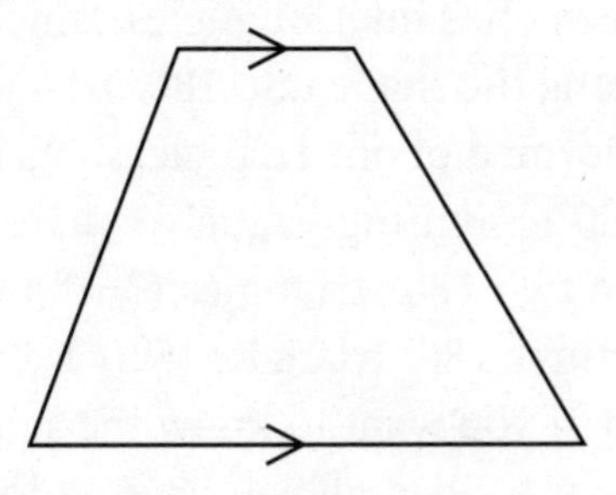

There are, of course, many other nameless quadrilaterals that have no equal sides or angles. This sad shape is an example:

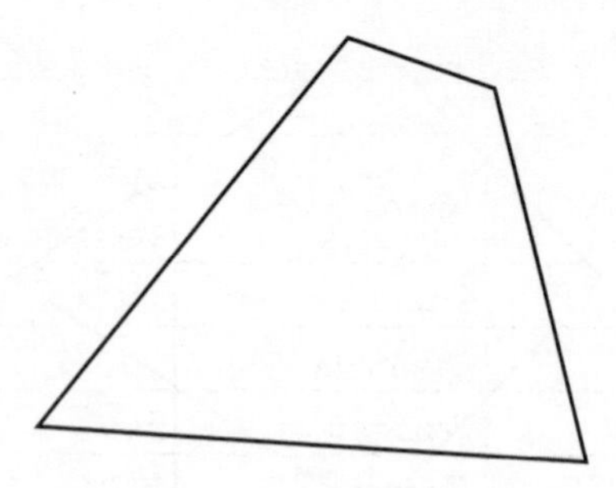

The one thing we *can* say about this shape is that, being a quadrilateral, its angles must all add up to 360°.

Never-Ending Polygons

Triangles and quadrilaterals are the straight-sided shapes we see the most of, but there are other polygons with any number of sides out there. The following chart shows some of the polygons you may have come across, with the sum of each one's interior angles, which we can work out by chopping the shape into 180° triangles. A triangle is obviously formed of one triangle, so its interior angles add up to 180°; anything vaguely square-based can be chopped into two 180° triangles (360°); pentagons are divided into three 180° triangles (540°); and so on.

In general, if you want to know the sum of angles for a shape that has n sides, all you have to do is take 2 from n and then multiply by 180°. This works for both regular and irregular polygons.

Sum of interior angles $= (n-2)\times 180^\circ$

Number of sides	**Name**	**Sum of interior angles**
3	Triangle	180°
4	Quadrilateral	360°
5	Pentagon	540°
6	Hexagon	720°
7	Heptagon	900°
8	Octagon	1080°
9	Nonagon	1260°
10	Decagon	1440°
12	Dodecagon	1800°
20	Icosagon	3240°

There are other polygons beyond this, each with an increasingly eccentric name.

EXTERIOR ANGLES

We know that the interior angle is the angle inside the corner of a polygon, so it would be terribly handy if the exterior angle were simply the whole angle outside the corner. Unfortunately, it's not.

Imagine you are walking around a polygon. The exterior angle is the angle you have to turn through at every corner – i.e the deviation from simply continuing in a straight line:

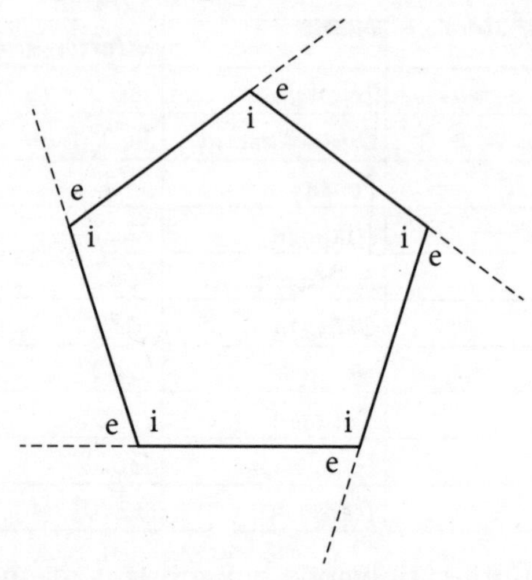

Every time you come to a corner of the pentagon, you deviate from a straight line at an angle equal to angle e.

As the diagram shows, the interior angle and the exterior angle always make 180°. If you know the size of the interior angle, you can simply deduct it from 180° to work out the exterior angle. This is easy with a regular pentagon, but an irregular shape is somewhat trickier:

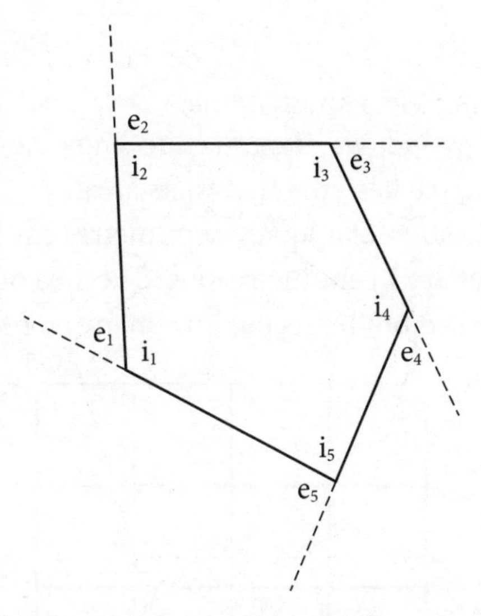

For both the regular and irregular pentagons, we're able to say that the sum of the interior angles is 540°. Can we say something similar about the exterior angles? Well, if you think about your walk around the polygon, a complete circuit will bring you back to where you started, facing in the same direction as before you set out. You will have turned 360°. This is true of any polygon, no matter how many sides it has.

Sum of exterior angles = 360°

AREA

In the real world, we often need to know how much flat space a shape takes up – that is, its area.

The definition of a square centimetre (cm^2) is a square whose sides are 1 centimetre long. The area of a rectangle can be worked out by seeing how many cm^2 fit inside it:

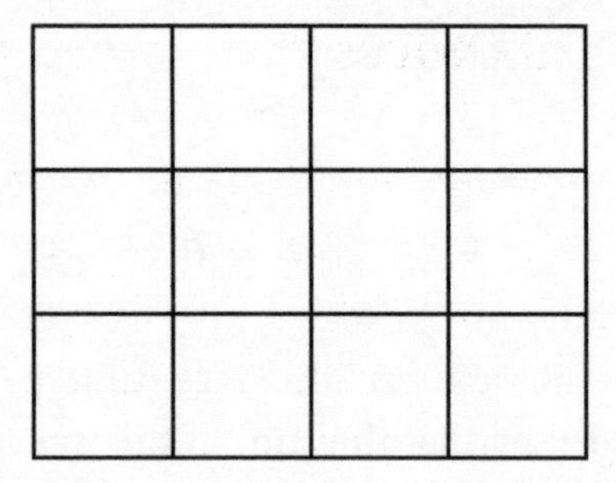

The area of this rectangle is $12cm^2$ because it contains twelve square centimetres. There are 3 rows of 4 squares, so $3 \times 4 = 12$.

This is useful as rectangles get bigger, but it also helps when they don't contain a whole number of squares:

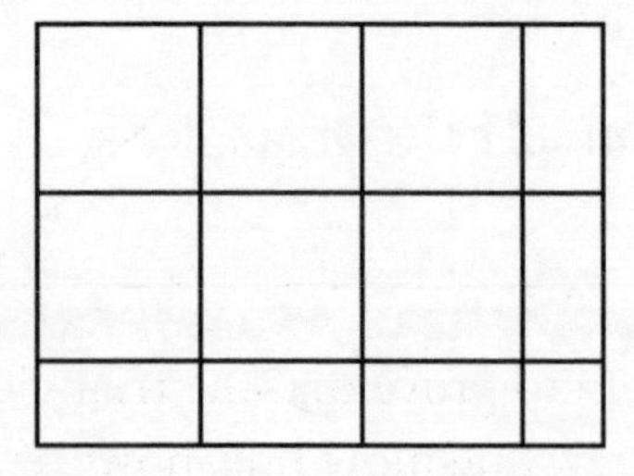

This rectangle has 2.5 rows of 3.5 squares, giving us 6 whole centimetre squares, 5 half centimetre squares and 1 quarter centimetre square, altogether giving an area of $6+(5\times0.5)+0.25=8.75\text{cm}^2$, which is the same answer we'd have got if we'd worked out 3.5×2.5.

Area of rectangle = length × width

Area and Triangles

I said earlier that any rectangle is two triangles put together. I'm going to go a bit further now and say that any triangle is half a rectangle. If you cut across a rectangle diagonally, you'll have two identical rectangles.

That theory works nicely with two right-angled triangles that can snuggle directly into the corners of a rectangle, but what if we have a non-right-angled triangle?

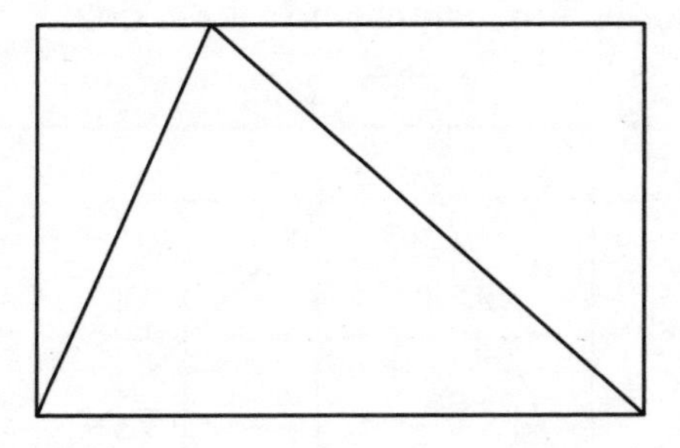

This doesn't seem to prove that the triangle is half of the rectangle. If I draw one more line, however:

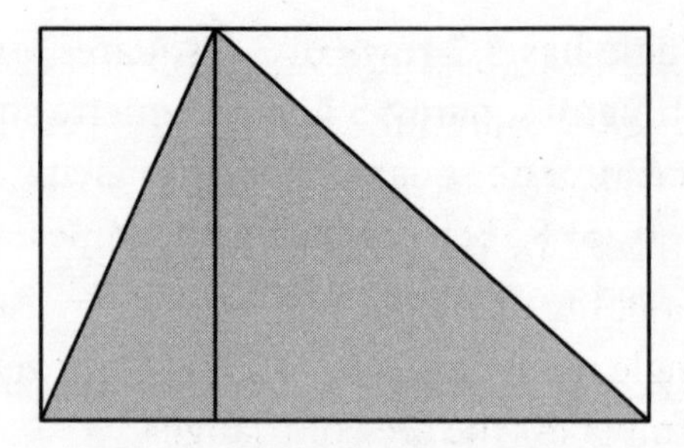

The large rectangle has been split into two smaller ones, each of which is half full of triangle, so it follows that the large rectangle must also be half full.

When we talk about the area of a rectangle, it's always in terms of length and width. With a triangle, however, we tend to refer to base and height. What often catches people out at this stage is that, most often, the height of a triangle doesn't correspond to the length of any of its sides. Think of a triangle as a mountain: its height above sea-level is not the same as the distance up the side of the mountain, but rather how vertically high the summit is. Let's add some dimensions and take a look:

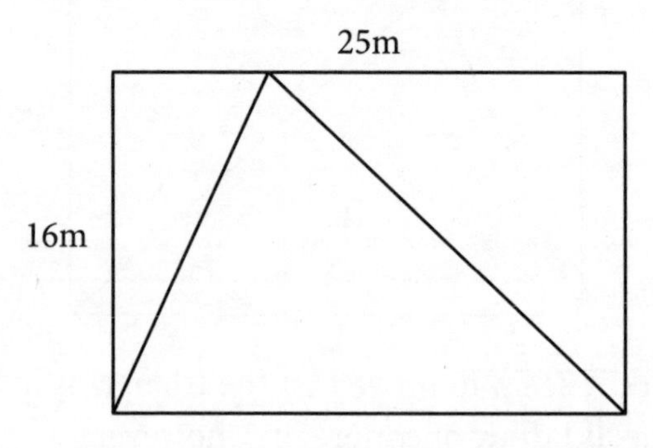

This rectangle has a base of 25 metres and a height of 16 metres.* In order to find the area of the rectangle – and thus of the triangle – we don't actually need to know the lengths of the triangle's other two sides.

The area of the rectangle is $25 \times 16 = 400\text{m}^2$, so the area of the triangle must be $400 \div 2 = 200\text{m}^2$. This leads us to a general formula for the area of triangles:

Area of triangle = (base × height)÷2
or the slightly snappier: ½ × base × height

Area and Other Shapes

The good news is that the odd-shaped polygons we've looked at can all be split into handy triangles.

Parallelograms are easily divided into two identical (isosceles) triangles:

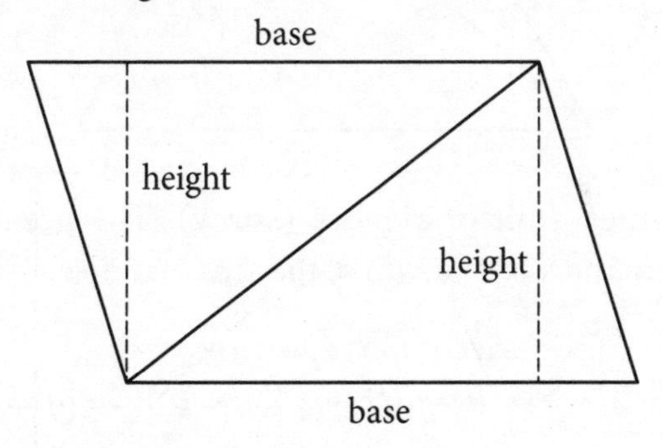

* Some people would say 'perpendicular height', to make it clear that they are referring to something that is at right angles to the base.

If the area of each triangle is ½ × base × height – which it is – the area of the whole parallelogram must be:

Area of parallelogram =
½ × base × height × 2
or indeed just: base × height

But be careful not to confuse the parallelogram's height with the length of one of its sides.

The way to chop a trapezium into triangles is a little less obvious, and you end up with two triangles of the same height, but with different bases. Traditionally these are labelled a and b:

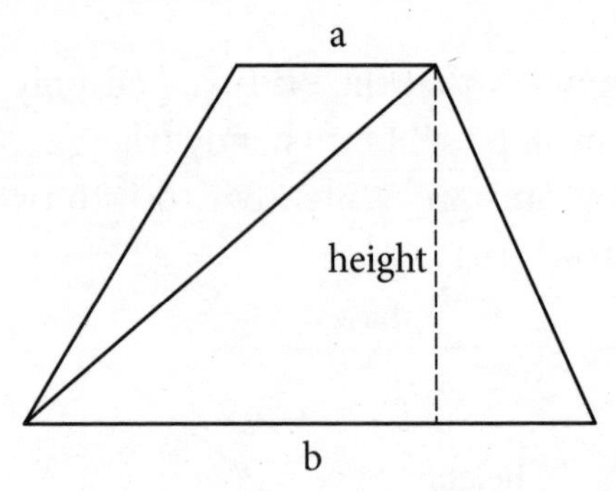

Now comes a bit of algebra (sorry). The area of the trapezium must be the sum of the area of the two triangles:

Area of trapezium =
(½ × a × height) + (½ × b × height)

To make things simpler, we can factorize this rather ungainly formula in exactly the same way as we did in the algebra chapter (see p.112). Both terms in the formula have a $1/2$ and a height in them, so they can go outside the brackets:

$$\text{Area of trapezium} = 1/2 \times \text{height} \times (...+...)$$

The only thing left to stick into the brackets is a and b:

$$\text{Area of trapezium} = 1/2 \times \text{height} \times (a+b)$$

This is often rearranged to look like this:

$$\text{Area of trapezium} = \frac{(a+b) \times h}{2}$$

In the same way that Pythagoras' Theorem has its wordy equivalent, some people find this easier to remember and use:

'The area of a trapezium is equal to the mean of the parallel sides multiplied by the distance between them.'

'The mean of the parallel sides' is the $(a+b)\div 2$ bit, and 'the distance between them' is h, the height.

Moving on, kites are very obviously two identical triangles that are back to back:

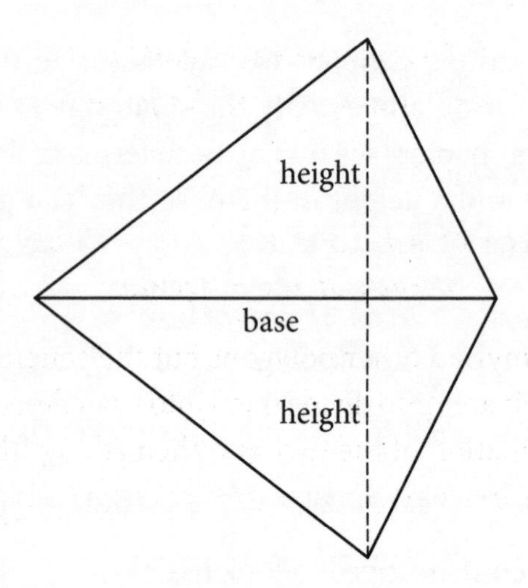

So we get:

$$\text{Area of kite} = (1/2 \times \text{base} \times \text{height}) + (1/2 \times \text{base} \times \text{height}) = \text{base} \times \text{height}$$

However, it seems a bit odd to talk about those distances as the base and the height of the kite. They are, in fact, its diagonals – diagonals being straight lines that connect one corner of a polygon to another non-adjacent corner. The base, as we have it, is one diagonal and the height is *half* the other diagonal, so we can say:

$$\text{Area of kite} = \text{base} \times \text{height}$$

(where base = diagonal 1 and height = half of diagonal 2)

$$\text{So area of kite} = \text{diagonal 1} \times \tfrac{1}{2} \times \text{diagonal 2}$$

Or, put more elegantly:
½ × diagonal 1 × diagonal 2

Or, in other words:

'The area of a kite is half the product of the length of its diagonals.'

There are myriad other polygons but the general rule for finding their area is to divide them into triangles, rectangles or a combination of the two, and then add up their areas to find the area of the shape you are considering.

CIRCLES

The triangle approach doesn't work at all for circles, however, since they famously have a curved side rather than multiple straight ones.

Technically speaking, a circle *is* a polygon, just one with so many minute straight sides that they blend into a smooth curve. The more sides a polygon has, the closer it comes to resembling a circle:

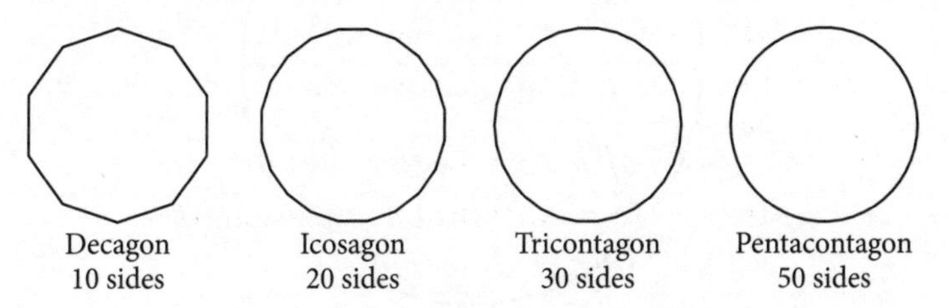

Circles have always occupied a special place in mathematics. Early scholars soon realized that the motion of planets and their ilk was curved like circles, which made them feel that the circle was somehow special and mystical. Of course, the wheel, widely believed to be one of humanity's most important inventions, also relies heavily on an understanding of circles.

There is lots of specialist vocabulary associated with circles:

- The perimeter (distance around the outside) of a circle is called the circumference.
- The distance from the centre of a circle to the edge is called the radius.
- The distance all the way across the circle, through the centre, is called the diameter. The diameter is effectively two radii (plural of radius).

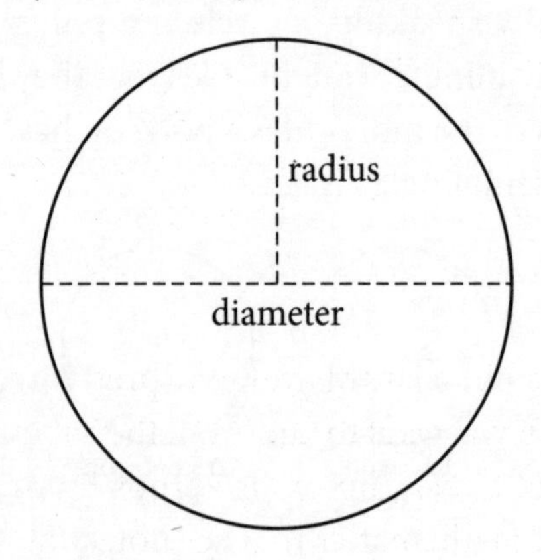

PI

Circles, like all regular polygons, have a special property: their proportions are all the same. All circles look identical; it is only their size that changes. This, in itself, is not particularly remarkable, especially to non-mathematicians. It does, however, have a useful by-product: as soon as there were circles, people started to notice that, if you divide the circumference by the diameter, you always get the same number…

The tricky part of dividing the circumference by the diameter is measuring the circumference. We're really good at measuring things that are straight, but measuring curved things is a bit more of a bother. So early mathematicians were only able to say that dividing the circumference by diameter produced roughly 3. This number is even mentioned in the Bible: a circular pond in King Solomon's temple is described as being three times as far around as it is across.

It turns out that this value is pretty important, especially if you want to figure out the circumference and area of circles, but it also crops up in all sorts of areas of mathematics that are not, at first glance, anything to do with circles.

It was discovered, mathematically, that the number was irrational – that is, it can't be expressed as an exact fraction, and as a decimal it goes on forever without repeating itself. Because infinitely long numbers can be a bit of a chore to write down, the number was given a symbol instead: π, pi, which is the Ancient Greek letter 'p'.

$$\pi = 3.14159265358979323846264338327 9\ldots$$

It's easy to think of pi as some sort of complicated entity, as many of my students have done, but it's crucial to remember that the handy symbol π just represents a number, and a very specific one at that.

Navigating Your Way Around Circles

The vocab associated with circles is a bit of a mouthful, so I'm going to use the following shorthand: c is circumference, r is radius and d is diameter. If I write down the discovery of π as a formula, I get:

$$c \div d = \pi$$

This is saying that the circumference of a circle divided by its diameter is π, i.e. three and a bit. If I multiply both sides of the formula by d, I get:

$$c = \pi \times d$$

This is very handy, because now, if I want to know the perimeter of a circle, I just need to know π – which is a fixed number that people with far too much time on their hands have worked out to trillions of decimal places – and d, the diameter, which is easy to measure with a ruler as it is straight.

If I can't be bothered to measure the whole diameter, I can simply measure the radius and double it. So now we have the two formulae that are drummed into every school child's head:

$$\text{Circumference of a circle} = \pi d = 2\pi r$$

The Area of a Circle

Finding the area of a circle is quite tricky if your way of finding an area involves cutting the shape up into triangles and rectangles.

But if you were to draw a circle with a radius of 1 metre and laboriously work out its area by counting how many square centimetres fit inside it, and then compare that number with the area of a square with 1-metre sides, you would find that the circle is 3.14159265358979323846 2643383279… ad infinitum times larger than the square. That's right: it's our old friend pi again.

Furthermore, it doesn't matter what size the circle is: if

its radius is r, it will always be π times bigger than a square with r length sides.

Area of a circle with radius r =
π × area of square with side r

The square, if its side is r long, will have an area of r×r:

Area of a circle with radius r = π×r×r
More commonly written as: πr^2

There are a couple of things to watch out for with this. The first is to make sure you square the radius, rather than just multiply by 2. The second is to be careful if you're using the diameter instead of the radius – it's all too tempting to square the diameter and divide by two, but in fact you have to divide by four:

$$r=\frac{d}{2}$$

$$r^2=(\frac{d}{2})^2$$

$$r^2=\frac{d}{2}\times\frac{d}{2}$$

$$r^2=\frac{d^2}{4}$$

SOLIDS

So far, we've only looked at two-dimensional shapes. But maths is often used as a way to describe and explain the world around us, which is abundantly three-dimensional, so we need to think about the 3D relatives of all the 2D shapes we've been dealing with.

PRISMS

The easiest way to go from 2D to 3D is to take any two dimensional shape and 'pull it out' into three dimensions.

I'll start off with a circle. If I lay a circle flat and pull it out in the direction of the arrow, I get a solid shape:

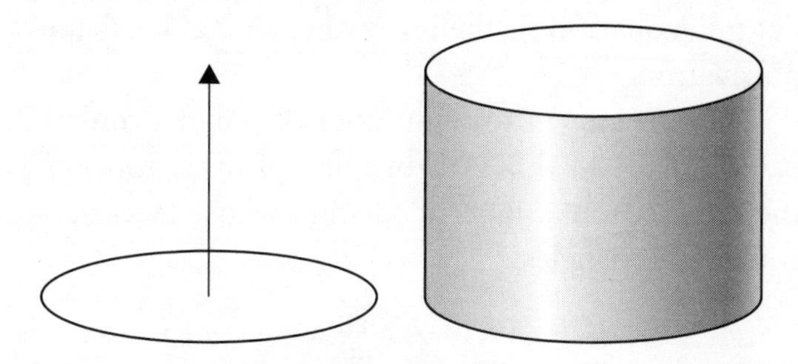

I could do this with any flat shape to obtain a solid shape. Shapes made in this way are called prisms. In this instance, the prism we have made is specifically a cylinder.

A prism made using a square or rectangle is a cuboid:

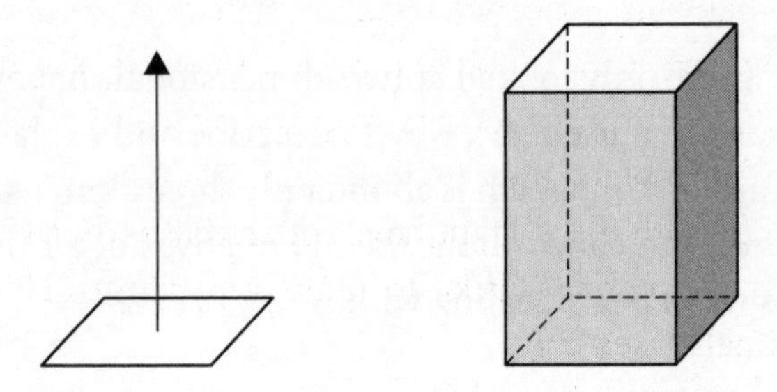

If my original shape had been square and I'd turned it into a cuboid whose faces were also all square, that would be a cube.

Polyhedrons and Cones

The second way of making solids is to take a number of flat polygons and fit them together into a 3D shape: a polyhedron.

There are an infinite number of polyhedrons that can be made in this way, but one of my favourites is the tetrahedron, a four-sided solid whose faces are all equilateral triangles:

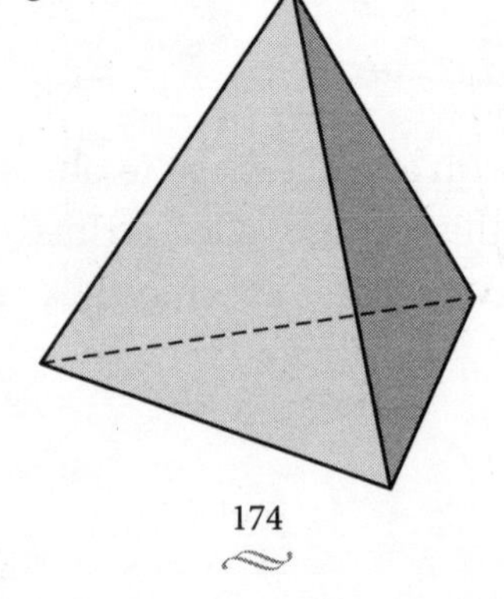

The tetrahedron is an example of a pyramid, which, to a mathematician, is formed by taking a polygon and, rather than pulling it up to make a prism, making everything meet at a point. So a tetrahedron is a pyramid that has a triangular base.

The pyramids in Egypt have square bases, but I could use any shape base I like to make a pyramid. Here's a hexagonal-based one:

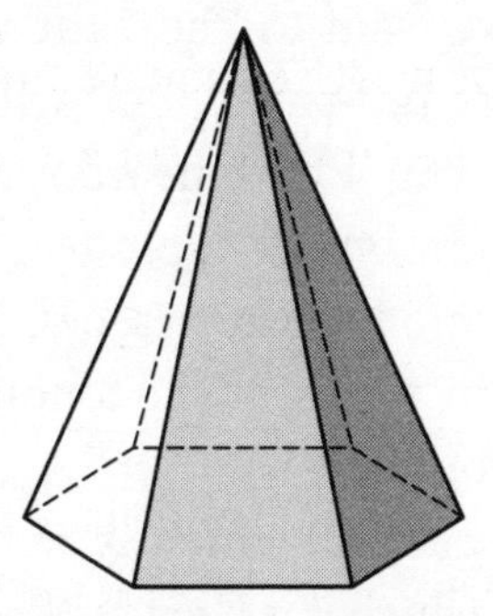

Another polyhedron that we are all used to seeing, whether we like it or not, is a football, which is made up of pentagons and hexagons in a solid called a truncated icosahedron:

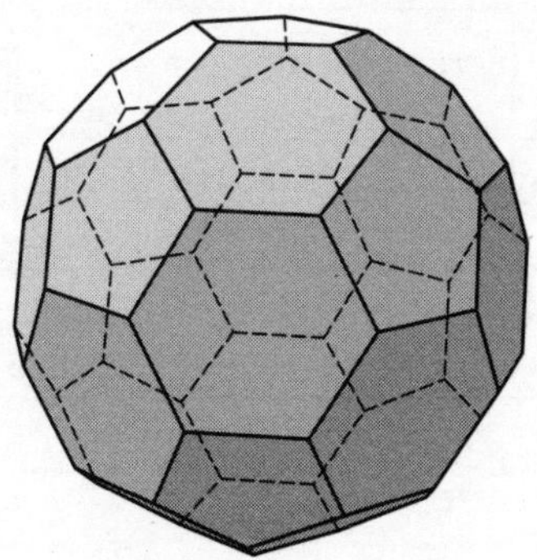

A circle can also be made into a pyramid-like solid called a cone, as well as a sphere – any golf ball or tennis ball is basically a sphere. A sphere is technically a polyhedron with an infinite number of tiny flat faces.

VOLUME

Volume can be measured using centimetres cubed. So it makes sense, if we want to know the volume of a cube or a cuboid, to see how many cm^3 will fit inside it. On a far larger scale, you'll have heard removal companies advertising the number of cubic feet of furniture and boxes they can fit into their lorries, or Olympic swimming pools described in terms of the amount of water they hold. It's all just volume.

Consider this cuboid, which is 5cm long, 2cm wide and 3cm tall, and which has been broken down into $1cm^3$ cubes:

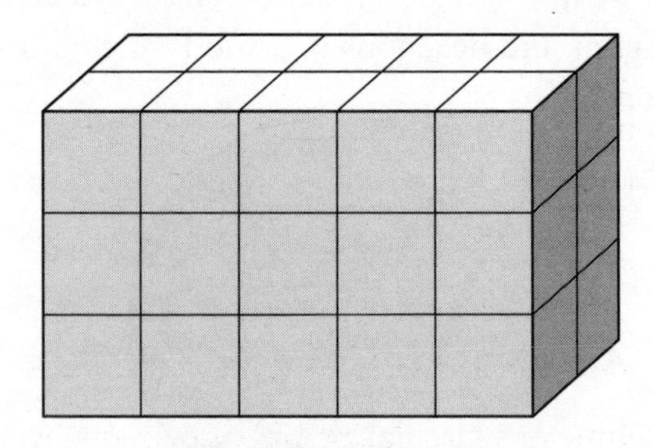

From here, I can simply count the number of cubes: 30. So the volume of the cuboid must be $30cm^3$.

If I were building this cuboid out of cubes, I'd need 10 cubes for the bottom layer because the bottom is 5cm long and 2cm wide. I'd then need 3 layers in total because the cuboid is 3cm tall. So the number of cubes I'll need is equal to the length, width and height of the cuboid multiplied together:

Volume of a cuboid = length × width × height

If the cuboid is a cube, then the length, width and height are all the same:

Volume of a cube = $length^3$

These formulae are nice and easy to remember, and can be exploited to find the volume of other prisms. To make a cuboid, we started with a rectangle and pulled it up to make a rectangular prism. The area of the original rectangle is length × width, which we then multiply by height to get the volume. It follows that, if you find the area of your starting shape and then multiply by the height to which you stretch it out, you can get the volume of the prism.

Volume of a prism =
Area of starting polygon × height

So the volume of a circular prism, or cylinder, would be:

$$\text{Volume of a cylinder} = \pi r^2 h$$

where r is the radius of the starting circle and h is the final height of the cylinder.

Cones and pyramids, being very similar, share the same volume formula. In the same way as the circumference of any circle divided by its diameter always gives the same number – π – it turns out that the volume of a cone or pyramid is always one third the volume of a prism with the same starting polygon:

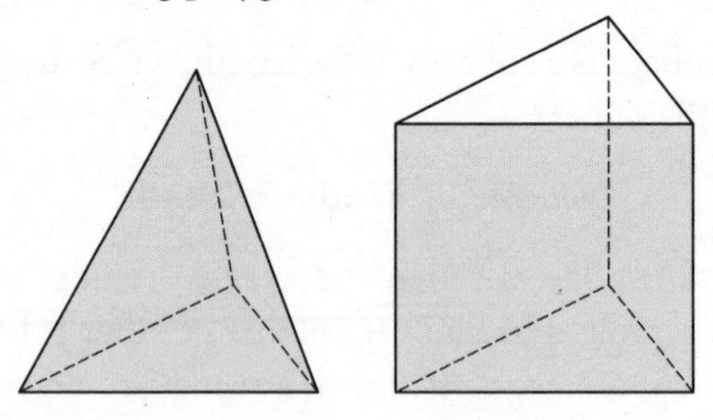

If the pyramid has a base area of B and a height of h, then:

$$\text{Volume of pyramid} = \frac{Bh}{3}$$

For a cone with a base circle of radius r:

$$\text{Volume of cone} = \frac{\pi r^2 h}{3}$$

The volume of a sphere is derived using calculus, a mathematical technique that I won't burden you with in this book, but I will just tell you the formula. You won't be surprised, I'm sure, to see that π is in there:

$$\text{Volume of a sphere} = \frac{4\pi r^3}{3}$$

TRIGONOMETRY

Trigonometry can basically be summed up as 'working out lengths and angles in triangles'. It demonstrates very nicely how hard it is to remember how to use something that you don't actually understand – it's rather like programming a video recorder. Trigonometry crops up in all sorts of areas of high-level pure and applied mathematics, and so school children are taught how to 'do' trigonometry without any real explanation of what on earth the actual point of it is.

The reason that the point of trigonometry is glossed over is that it's tricky and doesn't relate easily to everyday life. It could be said that David Beckham is a master of mathematics because he is able to solve complex equations of motion and fluid dynamics in his head to kick a ball in exactly the right place, at the right speed and at the right angle to bend it round the wall and beat the goalkeeper. But frankly I doubt he really troubles himself with the calculations: nobody does (or needs to do) trigonometry in their heads as part of the normal human experience. It's just the results of doing it that we need.*

In the past, explorers navigating the world's oceans

---

* I'm going to call trigonometry 'trig' from now on, to help save the environment.

used trig to find out where on Earth they were, calculating the angle between their location and the North Star, and working out a series of unknown angles and lengths. Nowadays, GPS makes all of that somewhat redundant, but architects and engineers still rely heavily on trig's rules and conventions.

MAKING RIGHT-ANGLED TRIANGLES

Humans love right angles. We want the walls of our houses to be at right angles to the floor. We like our shelves to be at right angles to the wall. Furniture, books, streetlights, buses: most things of a practical nature are formed out of straight lines and right angles. No wonder Pythagoras gave the right-angled triangle such special attention. This trig business picks up where Pythagoras' Theorem leaves off.

Put simply, his theorem says that if you know the lengths of any two sides of a right-angled triangle, you can figure out the third one. Trig expands upon this. If you know trig and you know

- one of the non-right angles and the length of one side; or
- the lengths of two of the sides,

trig lets you work out all the sides and angles of the triangle.

Magic? No: logic. Let's make a right angle out of two

lines. I'm going to go for a 3m line and a 4m line.

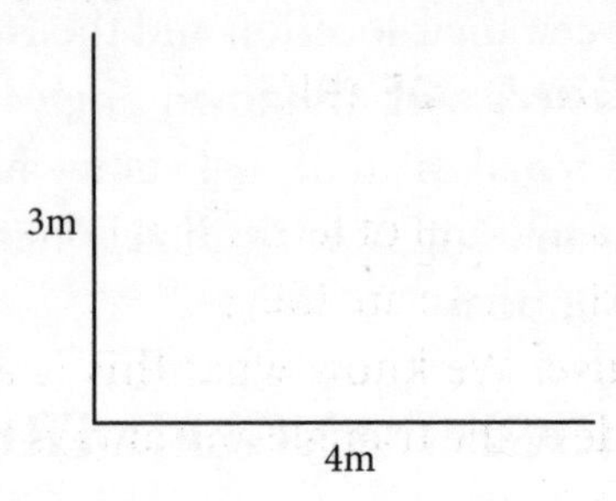

Now, here's the salient trig-related fact about this unassuming shape: the third side of my triangle has to be just the right length and at just the correct angle to fit in the gap I have. There's no other way to 'close' this triangle:

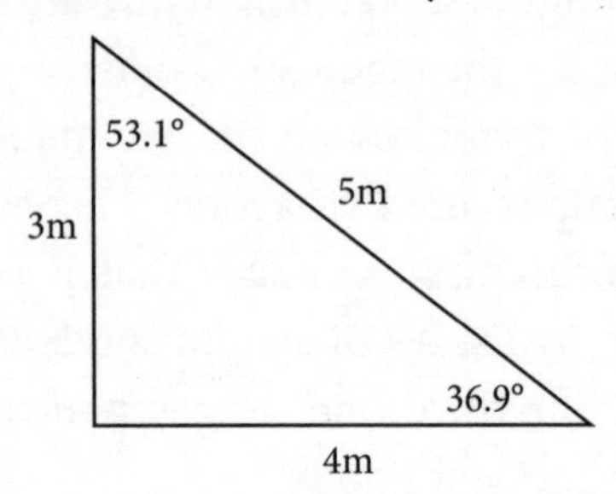

By choosing to have a 3m side and a 4m side that meet at a right angle, I've effectively 'set' the length of the other side and the size of the two angles as well.

THE LANGUAGE OF TRIG

There are a handful of terms that crop up time and again in trig. These are they:
Hypotenuse: We know what this is already: the longest side of the triangle and always opposite the right angle.
Opposite: The side opposite a non-right angle.
Adjacent: The side adjacent to, or touching, a non-right angle.

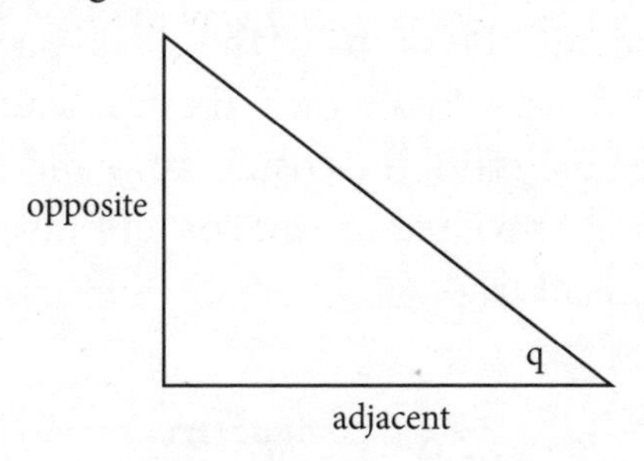

For angle q, the vertical side is opposite to it and the horizontal side is adjacent to it.

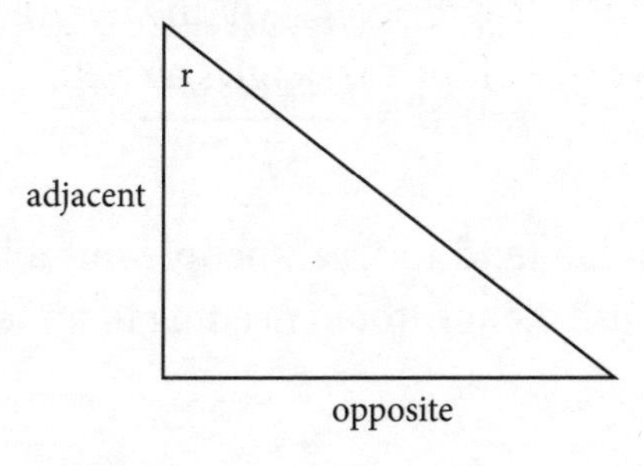

For angle r, the vertical side is adjacent and the horizontal side is opposite.

A side can be opposite or adjacent, depending on the angle in question.

SIN, COS AND TAN

Way back when, some very, very clever people noticed that there was a certain relationship between the angles in a triangle and the ratios of the sides – a ratio being one number divided by another. They came up with three ratios – the sine ratio, the cosine ratio and the tangent ratio (often abbreviated to sin, cos and tan), and these form the basis of trig.

The ratios are:

$$\sin \theta = \frac{\text{opposite}}{\text{hypotenuse}}$$

$$\cos \theta = \frac{\text{adjacent}}{\text{hypotenuse}}$$

$$\tan \theta = \frac{\text{opposite}}{\text{adjacent}}$$

These formulae tend to freak people out a lot, but you can happily use them without needing to understand how they are derived. The symbol θ (theta) is another Ancient Greek letter, often used in maths to stand for an angle.

Let's explore these ratios further using my original triangle:

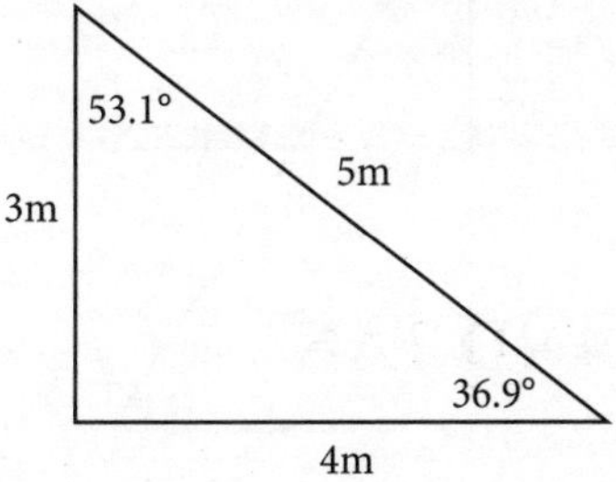

As far as the 53.1° in the top corner is concerned, the vertical 3m line is the adjacent side, the horizontal 4m is the opposite side and the 5m side is the hypotenuse. So if we used the sine ratio, we'd get:

$$\sin \theta = \frac{\text{opposite}}{\text{hypotenuse}}$$

$$\sin 53.1° = \frac{4}{5}$$

$$\sin 53.1° = 0.8$$

In the olden days, you'd have had to look up things like this in a big table but nowadays, thanks to the miracle of modern science, we have calculators and computers that can work these things out for us. If I put sin 53.1 into my calculator, I get 0.7996847,which is near as dammit 0.8.

So far, so good, but we haven't really seen how this can

be helpful. Well, let's look at a new triangle that's missing some info:

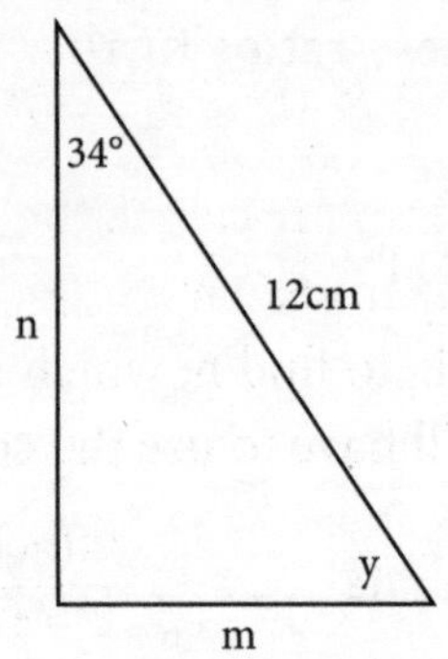

So now we don't know the lengths of two of the sides or the size of one of the angles. But we do have one side and one angle, so these other things, despite the fact that we don't actually know what they are yet, are already fixed: there is only one right-angled triangle in all of existence with a hypotenuse of 12cm and an angle of 34°.

Let's say I wanted to work out side m. I know an angle and the hypotenuse, and side m is opposite the one angle I do know, so I just have to work out which trig ratio has both opposite and hypotenuse in it – the sine ratio – and then use it:

$$\sin \theta = \frac{\text{opposite}}{\text{hypotenuse}}$$

$$\sin 34° = \frac{m}{12}$$

This is an equation. If I want to get m by itself, I need to multiply both sides by 12:

$$12 \times \sin 34° = m$$

If I put 12 × sin 34 into my calculator, I get 6.710315 – so I'll go ahead and say m=6.7cm.

I can also use trig to find n, which is the adjacent side to 34°. This time, I'll have to use the cosine ratio:

$$\cos \theta = \frac{adjacent}{hypotenuse}$$

$$\cos 34° = \frac{n}{12}$$

$$12 \times \cos 34° = n$$

$$n = 9.9cm$$

What about angle y? Well, ordinarily, you would exploit the fact that the angles in a triangle make 180°, so y must be 180–34–90=56°. But you can work it out with trig. Now that we know all of the sides, we can use any of the trig ratios. Let's plump for tangent as we haven't done that one yet:

$$\tan \theta = \frac{opposite}{adjacent}$$

$$\tan y = \frac{9.9}{6.7}$$

$$\tan y = 1.4776119$$

This is saying that the tangent ratio for y is 1.4776119. If I had a book of tables, I could work backwards to see which angle has that tangent ratio. Fortunately, my trusty calculator can do it for me, using a special manoeuvre called inverse tangent, the shorthand of which is $\tan^{-1}$:

$$\tan y = 1.4776119$$
$$y = \tan^{-1}(1.4776119)$$
$$y = 55.9°$$

Why the discrepancy between this answer and the one we worked out by subtracting the angles from 180°? Our values for the sides, 9.9cm and 6.7cm, were rounded and therefore not 100% accurate.

That's pretty much it for basic trigonometry – and it wasn't really that bad after all, was it?

INDEX